"The Object Lessons series achieves something very close to magic: the books take ordinary—even banal—objects and animate them with a rich history of invention, political struggle, science, and popular mythology. Filled with fascinating details and conveyed in sharp, accessible prose, the books make the everyday world come to life. Be warned: once you've read a few of these, you'll start walking around your house, picking up random objects, and musing aloud: 'I wonder what the story is behind this thing?'"

Steven Johnson, author of *Where Good Ideas Come From* and *How We Got to Now*

"Object Lessons describes themselves as 'short, beautiful books,' and to that, I'll say, amen. . . . If you read enough Object Lessons books, you'll fill your head with plenty of trivia to amaze and annoy your friends and loved ones—caution recommended on pontificating on the objects surrounding you. More importantly, though . . . they inspire us to take a second look at parts of the everyday that we've taken for granted. These are not so much lessons about the objects themselves, but opportunities for self-reflection and storytelling. They remind us that we are surrounded by a wondrous world, as long as we care to look."

John Warner, *The Chicago Tribune*

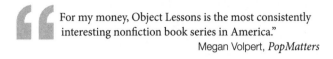
For my money, Object Lessons is the most consistently interesting nonfiction book series in America."
Megan Volpert, *PopMatters*

Besides being beautiful little hand-sized objects themselves, showcasing exceptional writing, the wonder of these books is that they exist at all . . . Uniformly excellent, engaging, thought-provoking, and informative."
Jennifer Bort Yacovissi, *Washington Independent Review of Books*

. . . edifying and entertaining . . . perfect for slipping in a pocket and pulling out when life is on hold."
Sarah Murdoch, *Toronto Star*

[W]itty, thought-provoking, and poetic . . . These little books are a page-flipper's dream."
John Timpane, *The Philadelphia Inquirer*

Though short, at roughly 25,000 words apiece, these books are anything but slight."
Marina Benjamin, *New Statesman*

OBJECTLESSONS

A book series about the hidden lives of ordinary things.

Series Editors:

Ian Bogost and Christopher Schaberg

In association with

BOOKS IN THE SERIES

sewer

JESSICA LEIGH HESTER

BLOOMSBURY ACADEMIC
NEW YORK • LONDON • OXFORD • NEW DELHI • SYDNEY

BLOOMSBURY ACADEMIC
Bloomsbury Publishing Inc
1385 Broadway, New York, NY 10018, USA
50 Bedford Square, London, WC1B 3DP, UK
29 Earlsfort Terrace, Dublin 2, Ireland

BLOOMSBURY, BLOOMSBURY ACADEMIC and the Diana logo are trademarks of
Bloomsbury Publishing Plc

First published in the United States of America 2023

Cover design: Alice Marwick

For legal purposes the Acknowledgments on p. 181 constitute an extension of this
copyright page.

Bloomsbury Publishing Inc does not have any control over, or responsibility for, any
third-party websites referred to or in this book. All internet addresses given in this
book were correct at the time of going to press. The author and publisher regret any
inconvenience caused if addresses have changed or sites have ceased to exist, but
can accept no responsibility for any such changes.

Library of Congress Cataloging-in-Publication Data
Names: Hester, Jessica Leigh, author.
Title: Sewer / Jessica Leigh Hester.
Description: 1st. | New York, NY: Bloomsbury Academic, 2022. | Series:
Object lessons | Includes bibliographical references and index. |
Summary: "Sewers are a mirror to the world above, at a time when our behaviors
are drastically reshaping the environment for the worse. Studying that murky warren
shows us who we truly are, and lays out a roadmap for being better stewards
of a quickly changing world"– Provided by publisher.
Identifiers: LCCN 2022019607 (print) | LCCN 2022019608 (ebook) |
ISBN 9781501379505 (paperback) | ISBN 9781501379512 (epub) |
ISBN 9781501379529 (pdf) | ISBN 9781501379536
Subjects: LCSH: Sewerage. | Sewage.
Classification: LCC TD673 .H47 2022 (print) | LCC TD673 (ebook) |
DDC 628/.2–dc23/eng/20220608
LC record available at https://lccn.loc.gov/2022019607
LC ebook record available at https://lccn.loc.gov/2022019608

ISBN: PB: 978-1-5013-7950-5
ePDF: 978-1-5013-7952-9
eBook: 978-1-5013-7951-2

Series: Object Lessons

Typeset by Deanta Global Publishing Services, Chennai, India
Printed and bound in the United States of America

To find out more about our authors and books visit www.bloomsbury.com and
sign up for our newsletters.

CONTENTS

INTRODUCTION

OUR SEWERS, OURSELVES

Why should we care about sewers, and what can they teach us?

The English town of Sidmouth is carpeted with emerald grass that ends at cliffs the color of burnt ochre. The steep, rugged rock faces look laminated, like pastry, and beneath them, waves froth on a pebbled beach. Those walls date back hundreds of millions of years. More recently, humans have scoured the shore for traces of ancient life, and used our hands and pickaxes to pry prehistoric jaws and fossilized footprints from the stone. Plucked from an area now nicknamed the Jurassic Coast, these treasures yield clues about how amphibians, reptiles, and other creatures ripped through flesh or fronds long before we walked the Earth.

In December 2018, an excavation of a different stripe was beginning near this storied shore. A crew descended into a sewer beneath a parking lot overlooking the water. At first, the underground world looked a bit like the inky innards of a cave—dark and damp. As the team spelunked the pipes, squelching through standing water and lighting their way with headlamps that bounced off their reflective vests, they passed heaps that looked like trampled stalagmites rising from the ground, or maybe piles of guano dropped by bats with overactive digestive systems.

They were neither. The guts of this place were blocked by a fatberg—a gloopy mound of oils, grease, and fats fastened in a matrix of wet wipes, hair, and other nasty stuff dumped down local sinks or flushed down toilets. The mess snaked 210 feet through the pipes, and the water company estimated that it would take several months to chip the thing apart. They could attack it with pickaxes or enlist high-powered jets of water to blast it into a chunky slurry, sucked up by vacuums feeding into trucks high above while workers slogged below.

It was the biggest fatberg the local utility company had ever encountered in the county of Devon, and the closest to the water, too. "I saw it and thought: 'What on earth?'" sewer worker Charlie Ewart told *The Guardian*. "It's really eerie in that bit of the sewer and it does look like something out of a horror scene, all congealed and glossy and matted together with all kinds of things."[1] By the time the clean-up crew finished in March 2019, more than 90,000 gallons of gunk had been hauled to the surface—36 truckloads filled with stuff no one had wanted to exist in the first place.[2]

The earliest sewers date back thousands of years. People built them in Mesopotamia, for instance, and on the island of Crete, where Minoans at the palace at Knossos used reserves of water to rinse waste from latrines and into a system of channels.[3] Ancient sewers weren't always great at ferrying feces away, improving flow, or sparing residents from stenches—sometimes the networks even spurred unsanitary side effects. Archaeologists have found fly pupae studding the remains of a sewer in Herculaneum, one of the Roman towns blanketed by ash and rocks spit out by Mount Vesuvius in the year 79. From a hygiene perspective, these networks may have backfired. Lacking the hairpin bends that now prevent sewer gases from drifting back up through the seat, early toilets gave flies a place to feast. "With easy access to human waste, flies could have transferred [fecal] matter and pathogens to people," wrote science journalist Chelsea Wald in a 2016 overview of ancient toilets for *Nature*. "Toilets and sewers and things didn't seem to improve the intestinal health of the Roman population," University of Cambridge paleopathologist Piers Mitchell told Wald. Mitchell suspects that an uptick in roundworm and whipworm from the Bronze Age to Iron Age and beyond may have been due to growers spreading human feces on crops. When people ate those poop-fed foods, parasite eggs were unwelcome garnishes.

But over the intervening millennia, sewers revolutionized many elements of sanitation. Disease eventually decreased—

as did stomach-roiling smells—when humans figured out how to efficiently wend waste below the streets instead of allowing it to pool on top of them in revolting rivulets of excrement.

Sewers don't exist everywhere, but where they do, the sunken conduits form a place's subterranean arteries. Large pipes intercept water that gathers in streets and rushes into grate-topped storm drains, and catch basins; smaller pipes fan out to homes as chutes for the waste flushed or poured down drains. (Stormwater and wastewater may or may not mingle.) The networks often lead to wastewater treatment facilities, where intrepid workers might wrangle the contents, sanitize them to varying degrees, and then turn some loose into bodies of water. In other cases—particularly when combined wastewater and stormwater sewers are overwhelmed by rain—sewage spurts into waterways raw, without any treatment whatsoever.

Things move most swiftly in the pipes when there's an influx of liquid, but even when the rush of water in a sewer trickles to a standstill, the tubes are dynamic environments. That makes sewers terrible places for aspiring fossils. Organic objects lucky enough to stick around for thousands or millions of years are preserved elsewhere, beneath sediment or tucked into bogs, places starved of oxygen and deprived of nutrients where there's little to hasten decay. Sewers, on the other hand, are bacterial orgies—no fun at all for something trying to linger as anything but an irredeemably unsavory sludge.

Unlike the bones lodged, eons ago, into the cliffs of the Jurassic Coast, the human wastes that enter the sewer—feces, of course, but also condoms, candy wrappers, wipes, and more—won't fossilize in these warrens. But they won't disappear, either.

It's easy to assume that the only alternative to sticking around forever is to be fleeting: to pass through then disappear, flush, woosh, goodbye. "People just assume things are taken care of," explained Stephanie Wear, a marine ecologist at The Nature Conservancy. "Flush it and forget it is what we want to do."

Maybe that's a relief, but it's not that simple. Nothing truly vanishes down in the pipes. Anything that winds up there makes it to the treatment phase, is spit out into a waterway, or gets lodged somewhere along the route—and wherever it hangs out, we can learn from it. We've got to.

* * *

Sewers' subterranean design obscures the scale, work, and smells to such a degree that it's hard to picture the journey from toilet bowl to treatment plant, from tube to tributary. The sunken conduits may invite dizzy, delicious speculation—who's to say there *aren't* teenage, crime-fighting turtles snarfing pizza in some cranny?—or utter inattention. For years, I sort of forgot about sewers, though I should have known better.

Growing up, I spent summers and weekends in a rural ranch house at the edge of a lake. We were at the mercy of the septic tank, and we tried to stay in its good graces. We didn't always manage. One summer, while my grandparents were visiting, the bowl wouldn't empty; it couldn't whisk away its holdings. My dad called someone to come out and look at the system and the pipe stretching beneath the yard. It was packed with dirt and striated with skinny white tree roots from a nearby evergreen. The setup had been installed when the house was built, then obscured for almost 30 years. That gave the roots plenty of time to infiltrate, and when they did, it got gross. As the crew dug up the yard to replace some parts, you could smell shit on the breeze. (And in the meantime, to use the toilet, we drove 10 miles to a donut shop in town.)

At the time, the sight and smell of the yard reminded me that sunken things like septic tanks and sewer pipes shouldn't be allowed to slip out of mind. But later, living in city apartments in Chicago and Brooklyn hooked up to municipal water, I only thought about the sewer when I reached for a plunger or when nose-prickling smells swirled around the street, flying up from a grate to greet me. I didn't dwell on the fragile septic system that once demanded my attention, or the sewers on which I had come to rely.

I'm not alone in shunting sewers to the back of my mind. Stephanie Wear of The Nature Conservancy now spends a lot of time thinking about sewer infrastructure; it comes with the territory as a co-founder of the Ocean Sewage Alliance, a collective of partner organizations working to minimize wastewater pollution in our oceans. But when she and I chatted about the networks, she realized they'd been fixtures in her life for longer than she remembered, and not just in the bathroom. Wear squinted, shook her head, and laughed, realizing, *whoa*: It was freshly dawning on her that she played in a sewer outfall pipe as a kid near Washington, D.C. The sometimes-slick pipe was a perfect place to wander in pursuit of crayfish and salamanders. "It was a bit smelly, but not enough to deter a kid," she told me. "You just stayed away when there was a lot of rain."

Salamanders and tube-tromping aside, I've heard versions of this a lot. During a Zoom tour of one of New York City's wastewater treatment plants, Alicia West, director of public design outreach at the city's Department of Environmental Protection, echoed the idea that people tend to overlook the pipes. "Most people flush their toilet and then don't think about it unless something comes back up at them," she said. But in an era when outdated sewer infrastructure is overwhelmed and under-equipped, and wetter weather drives more water down storm drains and more trash and sewage into waterways, that ignorance is no longer sustainable.

*   *   *

The notion that feces holds clues about habits and ailments isn't new. Archaeologists study ancient deuces to learn about past diets, trade routes, and the parasites that thrived in historic guts. Paleontologists and biologists mine mounds of guano or coprolites—fossilized poop—for revelations about which animals lived where, and what nourished them.[4] But studying sewage is a little different: If old poop can open a portal into the distant past, sewage—the stuff passing through the pipes right now—is a lens focused on the present.

Wastewater can show us what's going on inside people's bodies even before something is obviously awry. As COVID-19 has stampeded across the globe, researchers in dozens of countries have analyzed sewage for the RNA from SARS-CoV-2, the virus responsible for the illness.[5] Poop sleuths popped up around the planet: A research team led by Colleen Naughton, an assistant professor of civil and environmental engineering at the University of California, Merced, launched a digital dashboard—cheekily named COVIDPoops19—now listing more than 3,390 surveillance sites across 67 countries, including Japan, Iran, Kenya, and Chile.[6] Since people often shed the virus in feces, studying a place's wastewater is a way to identify asymptomatic cases, as well as people who wouldn't necessarily seek out treatment. "Not everyone is getting tested, but everyone is going to the bathroom," Gertjan Medema, a microbiologist at the KWR Water Research Institute in Nieuwegein, the Netherlands, told *Nature*.[7]

Researchers can sometimes detect even one infection per 100,000 people. In addition to studying the wettest part of wastewater, scientists also examine the solid clumps within it. Since 2020, Alexandria Boehm, an environmental engineering professor at Stanford, and her team have been surveying the solid waste that floats through California's sewersheds each day. Boehm and her collaborators believe the fecal solids are even more valuable than their sloshier liquid surroundings.

Testing sewage can help scientists predict, with a few days' notice, where outbreaks are poised to get gnarly. Based on concentrations of SARS-CoV-2 in wastewater, a team of researchers studying Massachusetts concluded that telltale signs of infection first showed up in local sewers on March 3, 2020, before many clinical cases appeared locally.[8] A team in the Netherlands detected the RNA in sewage in the city of Amersfoort six days before cases came to light.[9] Since 2020, COVIDPoops19 has doubled the number of countries on the dashboard, and quadrupled the number of universities. (Schools have taken different approaches to spying on students' stool: Some dispatched researchers suited up in PPE down into labyrinthine tunnels to tap into pipes, NPR reported, while others relied on pumps to collect samples every 15 minutes in 24-hour blitzes.)[10] Two years into the pandemic, "there are still places adding more wastewater testing, especially with Omicron," Naughton told me in early 2022, as the Centers for Disease Control and Prevention geared up to add sewage monitoring results to its public-facing COVID data site. Sewage might have started to look especially shiny and appealing to public health experts as patients struggled to find PCR appointments and stores ran out of rapid tests for at-home swabbing. Communities were "flying blind at times," Boehm explained in a webinar about

SARS-CoV-2 in poop, hosted by the International Water Association. "And wastewater is always there, every single day, to help them see what's happening."

And it's a multi-purpose tool. Epidemiologists can track the viral load in wastewater to gauge whether cases are rising or declining in a particular place, and because each variant—Delta, Omicron, whatever comes next—leaves its own signature, scientists can sift for evidence that one is beginning to elbow others out. After researchers found 12 Omicron-associated mutations in New York City wastewater samples from November 2021, they reckoned that the variant may have already been passing through the city's pipes before it had even been announced to the world, and more than a week before America's first clinically confirmed case. (They couldn't say for sure: The virus in the sample was so fractured that scientists weren't able to confirm that the mutations they observed were all causing a ruckus on a single genome.[11] Researchers have also encountered viral fragments with an array of "cryptic" mutations they haven't pegged to any particular variant, leaving scientists wondering whether these versions might be tearing through rats, skunks, or other non-human creatures roaming the city.)[12]

Sewage undergoes the "exact same test" humans experience when we swivel swabs around our nostrils, Naughton told me—but wastewater is "a more complex and messier matrix" than a single person's schnozz. But these efforts pay off by informing public health interventions. By "snooping on sewage," as Sabrina Imbler and Emily Anthes put it in the *New York Times*, officials in Houston, Texas determined which ZIP codes might benefit from a targeted vaccine push or more testing supplies.[13]

Still, this type of surveillance doesn't work everywhere. Sewers are a barometer of privilege. Many parts of the world lack elaborate sewer systems, or any at all, and even in wealthy countries, service can be a crapshoot: America, for instance, has a patchwork of sanitation systems, relatively reliable in some places and nonexistent in others.[14] Because there's no way to efficiently test scattered septic tanks or snapshot communities with splintered municipal systems, rural and impoverished areas fall through the cracks, an environmental scientist in Jaipur, India, explained to *Nature*. Wastewater monitoring is more common in high-income countries, "similar to global vaccination disparities," Naughton pointed out, with wealthy countries distributing more jabs and surveying more wastewater than less-resourced ones.

Even in places where most residents do have sewer service and it's possible to study sewage on a large scale, fecal testing doesn't detect every case of COVID-19: Though many scientists have seen a pretty stable relationship between indications of infection in stool or wastewater and clinical cases in their communities, one study found that, of 42 patients with confirmed cases of the illness, only 28 had the viral RNA in their stool samples.[15] But sewage testing has proved to be helpful in places where dense clusters of people poop. That includes schools, but also airplanes, where people take care of business in close, confined quarters. Wastewater sampled from 37 long-haul flights to Australia was a good predictor of whether passengers would test positive during their subsequent mandatory 14-day quarantine, even though all fliers over the age of 5 had received negative tests in the two days before boarding.[16]

Wastewater can't pinpoint exactly who has the virus, but it can show scientists where to look. In August 2020, the University of Arizona quashed a potential outbreak by identifying and isolating two asymptomatic students.[17] When wastewater from one dorm came back positive, all of the residents were tested for COVID-19, and the infected poopers were quarantined. (Later, researchers studying campus wastewater would estimate that scrutinizing a college's sewage for signs of COVID-19 had a "predictive power similar to that of random clinical testing," at less than 2 percent of the cost.[18])

The broader field of wastewater epidemiology—surveilling sewage to learn about human behavior, disease, and the substances we ingest—predates COVID-19 by decades. One of its US champions is Christian Daughton, a former environmental chemist at the Environmental Protection Agency and a longtime advocate of analyzing sewage for pharmaceuticals, controlled substances, and other compounds that often survive wastewater treatment plants intact. In 2001, he described his idea to sample for drugs at wastewater treatment facilities as "a rare bridge between the environmental and social sciences," and it was and continues to be.[19] Several hundred samples, many collected by the

EPA in national surveys at wastewater treatment facilities in 2001 and 2006-7, live in the National Sewage Sludge Repository at Arizona State University, where environmental health engineer Rolf Halden and his lab members stack the plastic containers neatly in refrigerators, as if they're jars of condiments. Researchers use the samples to, for instance, better understand the chemicals we encounter in consumer products. "This work lets us put a finger on the chemical pulse of a nation," Halden told science reporter Carrie Arnold for a piece in the journal *Environmental Health Perspectives*.[20] But as journalist Miranda Weiss recounted in a long article about Daughton's work in the magazine *Undark*, these proposals have been controversial.[21] Budgets were slashed over concerns about privacy, even though poop (and whatever secrets it held) couldn't be traced back to a specific butt or its owner.

But the idea has caught on elsewhere, including Europe, Australia, and China. Scientists have looked to sewage to understand whether cocaine use spikes on the weekends—in some places, it does—and to gauge changes in community health.[22] On the heels of Greece's crushing debt crisis, which led to a major uptick in unemployment, a team of researchers found an increase in antidepressants, antipsychotics, benzodiazepines, and recreational drugs such as MDMA and methamphetamine in the wastewater of Athens.[23] The researchers blamed the knotted problems of declining investments in public health and "severe socioeconomic changes." Proving Daughton's point, wastewater laid bare the unfortunate intersection of the two. Beyond SARS-CoV-2, wastewater can also help researchers keep an eye on outbreaks of dengue virus, yellow fever, and Zika, plus hepatitis A and noroviruses (pathogens sometimes responsible for gut-emptying vomiting and diarrhea).[24]

Studying wastewater is also a way to cut through the bullshit. In 2005, Italian researchers sampling wastewater treatment plants and the River Po found the equivalent of nearly nine pounds of cocaine in the water, which they calculated as about 40,000 doses a day—far above official estimates of local drug use.[25] The team argued that sewage doesn't lie. It's pretty easy to avoid fessing up to a behavior in a survey about your habits; it's a lot harder to fudge something in poop.

Feces isn't the only stuff of anthropological interest in the sewers. Trash that winds up at treatment facilities is evidence of what people buy and toss—and the garbage corralled by contraptions at entrances to the sewers can be informative, too. The New Zealand-based company Enviropod manufactures a device called LittaTrap, which looks like a mesh deep-fryer and sits inside catch basins to intercept large debris, essentially a "trash can in a drain," explained Darren Tiddy, the former technical lead. The team analyzes the detritus that collects in those basins—which need to be emptied several times a year—for a partial portrait of the people who live, work, commute, or revel near the 50,000 traps installed in Australia, America, and beyond. In one instance, when a trap was strewn with streamers, paper, and colorful plastic, "we could literally link them back to a graduation party at a school," Tiddy said.

The trash that makes it down into the pipes also lends itself to forensic analysis. A lot is intercepted in the process of treating sewage, but the jumble that isn't eventually makes its way to streams, lakes, and oceans. When sewers pour unsavory cocktails of waste and garbage into those ecosystems, pipes smudge human fingerprints across landscapes that would be better off without them.

Bits of trash don't stay put where we dump them. Buoyed and shuttled by sewers, waves, and wind, they rove to places that the discarder may have never been. Plastics can enter the water and atmosphere in several ways, including breaking off litter, boats, and fishing gear—but they also wend their way there from our streets or homes, through foul rivers carrying water from our toilets, sinks, washing machines, and showers. (More on this in Chapter 4.) The stuff we buy, cast off, and flush can really muck things up, and the legacy of these habits may be visible to those who come later.

Though wet wipes and garbage won't loiter in the sewer for millions of years, the miniscule fibers and other plastics that slough off of them and tumble into the water will endure on shores or in sediments for generations. However plastic enters an ecosystem, it's persistent once it gets there: Scientists in California have already found some scraps dating back 50 years embedded in marine sediment off the coast of Santa Barbara.[26] The confetti-sized flakes were sandwiched between organic materials such as fish scales and plankton exoskeletons that had drifted to the sandy bottom. Some microplastics also fly up from sea spray—and because they aren't very dense, they are easily lofted high into the air and carried vast distances by wind. Janice Brahney, an environmental biogeochemist at Utah State University, and her collaborators recently suggested that microplastics small enough to be ferried as aerosols can hitchhike through the air, zooming from one continent to another.[27] These bits eventually fall to Earth, dusting even remote, landlocked, and preserved places such as national parks with synthetic snow.

The sewer is a trash chute, jumbling up our discarded waste with the natural things that have been growing, dying, and decaying for as long as there have been oceans and rocks, and creatures living alongside them.

*   *   *

Tell someone you're writing a book about sewers, and you're likely to be met with confused glances or nervous chuckles. Many people consider the pipes either impolite topics of conversation—the stuff of scatological jokes—or compelling curios, full of intrigue as well as waste. "It has been said that sewers exercise a curious fascination upon otherwise healthy and happy people," writes novelist and biographer Peter Ackroyd in his nonfiction account *London Under: The Secret History Beneath the Streets*.[28] The tunnels tug on those who relish the macabre, titillating, or unfamiliar, Ackroyd adds, with the laypeople keen to descend playing the role of "tourists seeking sensations."

Ackryod likens descending and wandering a sewer to the experience of entering the underworld—of Orpheus looking to fetch Eurydice, or of Virgil guiding Dante through Hell. To venture into the sewer, he suggests, is to visit an alien land, and then hopefully return home again with stories to tell.

But while the sewer is physically separate from the surface, it's not a realm distinct from ours. It's a place we have built—one we can study to learn about ourselves, and one we ought to care about.

In his lyrical book *Underland: A Deep Time Journey*, a brainy travelogue chronicling treks into crevices and crannies, writer Robert Macfarlane observes that across time and geographies, humans have enlisted underground spaces for three particular purposes. We have used these nooks, he writes, "to shelter what is precious, to yield what is valuable, and to dispose of what is harmful."[29] The sewer has historically been a place to dump and expel things we don't want. Increasingly, scientists are seeing it or its contents as a source of both energy and insights. The sewer is precious in its own right, and precarious.

We need to understand it. Studying the underground highways from homes and businesses to the bigger, natural world beyond them is a way to keep an eye on the windshield and the historical rearview mirror, all at once. The sewer is a window into the goods we buy and discard, the design and management of cities, what we cook and eat, what makes us sick, and how we are changing environments far beyond our toilets, flushing our influence from one place to another. Look closely, and you'll find that sewers—and the stuff that travels through them—tell us a lot about how humans live, what we're up to, and where we're going. The muck shows us a picture of who we are, and how we're reshaping the world.

Notes

1 Steven Morris, "'Horror Scene': Meet the Man Who Found the Sidmouth Fatberg," *The Guardian*, January 11, 2019, https://www.theguardian.com/environment/2019/jan/11/horror-scene-the-man-who-found-the-sidmouth-fatberg.

2 Jessica Leigh Hester, "A Pretty, Seaside Town in England Vanquished Its Giant Fatberg," *Atlas Obscura*, April 1, 2019, https://www.atlasobscura.com/articles/sidmouth-devon-england-fatberg.

3 Chelsea Wald, "The Secret History of Ancient Toilets," *Nature* 533 (May 26, 2016): 456–58, https://doi.org/10.1038/533456a.

4 Researchers study all sorts of ancient poop. See, for instance, this fun paper about guano: L. R. Gallant et al., "A 4,300-year History of Dietary Changes in a Bat Roost Determined From a Tropical Guano Deposit," *JGR Biogeosciences* 126, no. 4 (March 2021), https://doi.org/10.1029/2020jg006026.

5 "National Wastewater Surveillance System (NWSS)," CDC, last reviewed March 21, 2022, https://www.cdc.gov/healthywater/surveillance/wastewater-surveillance/wastewater-surveillance.html.

6 Colleen C. Naughton et al., "Show us the Data: Global COVID-19 Wastewater Monitoring Efforts, Equity, and Gaps," *medRxiv*, November 28, 2021, https://doi.org/10.1101/2021.03.14.21253564. The COVIDPoops19 dashboard is accessible at https://www.arcgis.com/apps/dashboards/c778145ea5bb4daeb58d31afee389082.

7 Freda Kreier, "The Myriad Ways Sewage Surveillance Is Helping Fight COVID Around the World," *Nature*, May 10, 2021, https://doi.org/10.1038/d41586-021-01234-1.

8 Fuqing Wu et al., "SARS-CoV-2 RNA Concentrations in Wastewater Foreshadow Dynamics and Clinical Presentation of New COVID-19 Cases," *Science of the Total Environment* 805 (January 2022), https://doi.org/10.1016/j.scitotenv.2021.150121.

9 Gertjan Medema et al., "Presence of SARS-Coronavirus-2 RNA in Sewage and Correlation With Reported COVID-19 Prevalence in the Early Stage of the Epidemic in The Netherlands," *Environmental Science & Technology Letters* 7, no. 7 (May 2020): 511–16, https://doi.org/10.1021/acs.estlett.0c00357.

10 Elissa Nadworny, "Colleges Turn to Wastewater Testing in an Effort to Flush Out the Coronavirus," NPR, October 26, 2020, https://www.npr.org/2020/10/26/925831847/colleges-turn-to-wastewater-testing-in-an-effort-to-flush-out-the-coronavirus.

11 Amy E. Kirby et al., "Notes From the Field: Early Evidence of the SARS-CoV-2 B.1.1.529 (Omicron) Variant in Community Wastewater—United States, November–December 2021," Centers for Disease Control and Prevention *Morbidity and Mortality Weekly Report*, 71 no. 3 (January 21, 2022): 103–105, http://dx.doi.org/10.15585

/mmwr.mm7103a5. See also: Emily Anthes, "Omicron Was Probably in N.Y.C. Well Before the First U.S. Case Was Detected, Wastewater Data Suggest," *New York Times*, January 20, 2022, https://www.nytimes.com/2022/01/20/health/nyc-omicron-wasteweater.html.

12 Davida S. Smyth et al., "Tracking Cryptic SARS-CoV-2 Lineages Detected in NYC Wastewater," *Nature Communications* 13 (February 2022), https://doi.org/10.1038/s41467-022-28246-3.

13 Emily Anthes and Sabrina Imbler, "In Sewage, Clues to Omicron's Surge," *New York Times*, January 19, 2022, https://www.nytimes.com/2022/01/19/health/covid-omicron-wastewater-sewage.html.

14 Researchers recently found that roughly 220,300 households in America's 50 largest metropolitan areas lived without piped water for drinking or flushing, and that this "plumbing poverty" strongly correlated with race and ownership status of one's home. The researchers noted that households helmed by people of color were 35 percent likelier to be "unplumbed" compared to homes headed by white residents; renters were also much more likely to go without. Communities of color also often wait a long time for sewer repairs, an experience some experts have classified as a type of environmental racism. See, for instance: Katie Van Syckle, "Raw Sewage Flooded Their Homes. They're Still Waiting for Help," *New York Times*, April 15, 2021, https://www.nytimes.com/2021/04/15/nyregion/sewage-pipe-flood-queens.html.

For more on plumbing and social and environmental justice, see: Catherine Coleman Flowers, *Waste: One Woman's Fight Against America's Dirty Secret* (New York: The New Press, 2020).

15 Yifei Chen et al., "The Presence of SARS-CoV-2 RNA in the Feces of COVID-19 Patients," *Journal of Medical Virology* 92, no. 7 (July 2020): 833–40, https://doi.org/10.1002/jmv.25825.

16 These samples only tell a partial story, the researchers add—not everyone uses the bathrooms aboard the plane. See: Warish Ahmed et al., "Wastewater Surveillance Demonstrates High Predictive Value for COVID-19 Infection on Board Repatriation Flights to Australia," *Environment International* 158 (January 2022), https://doi.org/10.1016/j.envint.2021 .106938.

17 Jaclyn Peiser, "The University of Arizona Says It Caught a Dorm's COVID-19 Outbreak Before It Started. Its Secret Weapon: Poop," *Washington Post*, August 28, 2020, https:// www.washingtonpost.com/nation/2020/08/28/arizona -coronavirus-wastewater-testing/.

18 Jillian Wright et al., "Comparison of High-Frequency In-Pipe SARS-CoV-2 Wastewater-Based Surveillance to Concurrent COVID-19 Random Clinical Testing on a Public U.S. University Campus," *Science of the Total Environment* 820 (May 2022), https://doi.org/10.1016/j.scitotenv.2021.152877.

19 Christian G. Daughton, "Illicit Drugs in Municipal Sewage: Proposed New Nonintrusive Tool to Heighten Public Awareness of Societal Use of Illicit-Abused Drugs and Their Potential for Ecological Consequences" in *Pharmaceuticals and Personal Care Products in the Environment: Scientific and Regulatory Issues*, ACS Symposium Series, vol. 791 (July 2001): 348–64, https://doi.org/10.1021/bk-2001-0791.ch020.

20 Carrie Arnold, "Pipe Dreams: Tapping Into the Health Information in Our Sewers," *Environmental Health Perspectives* 124, no. 5 (May 2016): A86–91, https://doi.org/10 .1289/ehp.124-A86.

21 Miranda Weiss, "In the Tales Told by Sewage, Public Health and Privacy Collide," *Undark*, April 21, 2021, https://undark .org/2021/04/21/covid-19-data-down-the-drain/.

22 Billy Hunter, "The Drugs Which End up in Our Sewers, Drains and Rivers," *RTÉ Brainstorm,* February 10, 2020, https://www.rte.ie/brainstorm/2020/0205/1113343-the-drugs -which-end-up-in-our-sewers-drains-and-rivers/.

23 Nikolaos S. Thomaidis et al., "Reflection of Socioeconomic Changes in Wastewater: Licit and Illicit Drug Use Patterns," *Environmental Science & Technology* 50, no. 18 (August 2016): 10065–72, https://doi.org/10.1021/acs.est.6b02417.

24 Waqar Ali et al., "Occurrence of Various Viruses and Recent Evidence of SARS-CoV-2 in Wastewater Systems," *Journal of Hazardous Materials* 414 (July 2021), https://doi.org/10 .1016/j.jhazmat.2021.125439. See also: Franciscus Chandra et al., "Persistence of Dengue (Serotypes 2 and 3), Zika, Yellow Fever, and Murine Hepatitis Virus RNA in Untreated Wastewater," *Environmental Science & Technology Letters* 8, no. 9 (August 2021): 785–791, https://doi.org/10.1021/acs .estlett.1c00517.

25 Ettore Zuccato et al., "Cocaine in Surface Waters: A New Evidence-Based Tool to Monitor Community Drug Abuse," *Environmental Health* 4 (August 2005), https://doi.org/10 .1186/1476-069X-4-14. For more in this vein, check out: Eric Hagerman, "Your Sewer on Drugs," *Popular Science*, February 22, 2008, https://www.popsci.com/scitech/article/2008-02/ your-sewer-drugs/.

26 Jennifer A. Brandon, William Jones, and Mark D. Ohman, "Multidecadal Increase in Plastic Particles in Coastal Ocean Sediments," *Science Advances* 5, no. 9 (September 2019), https://doi.org/10.1126/sciadv.aax0587.

27 Janice Brahney et al., "Constraining the Atmospheric Limb of the Plastic Cycle," *Proceedings of the National Academy of Sciences* 118, no. 16 (April 2021), https://doi.org/10.1073/pnas.2020719118.

28 Peter Ackroyd, *London Under: The Secret History Beneath the Streets* (New York: Anchor Books, 2012), 93.

29 Robert Macfarlane, *Underland: A Deep Time Journey* (New York: W. W. Norton & Company, 2019), 8.

1 CATHEDRALS OF SEWAGE

Sewers and wastewater treatment works can be expressions of civic ambition and pride.

The prince's presence had been eagerly anticipated. In April 1865, a steamer carried the Prince of Wales along the Thames, traveling east from his royal residence to Crossness. He was headed for a quiet, marshy shore some 12 miles downstream of London. When the boat docked on one of the river's southern banks, all who disembarked padded along a pier cloaked in crimson cloth. As the coterie strolled past shrubs and potted flowers, ushered along by the Royal Marines' band, someone raised a Royal standard. Other banners flapped from surrounding buildings.

The prince wasn't there to glad-hand with dignitaries or wave from the midst of a parade. Along with the Duke of Cambridge, the Archbishop of Canterbury, the Archbishop of York, and a smattering of legislators and well-heeled men, he had journeyed to Crossness to celebrate the sewer.

The facility hadn't yet gurgled into action; this was its birthday. Before the sludge started moving—the big event, the infrastructural equivalent of blowing out the candles on a cake—the visitors got a lay of the land. Guests roamed the engines, boiler house, and culvert, which would store sewage that had passed through the quartet of massive pumps. Without any muck sloshing through the space, noted a reporter for *The Illustrated London News*, "There was nothing about it to remind one of the purpose for which it had been built. It was simply a long, lofty, wide tunnel of excellent brickwork."[1] It was even beautiful, or something close to it, managing "a cheerful appearance."

The group then pressed on to the reservoir, an "immense range of vaults" lit so that "the architectural features of the works and the nicety with which the brickwork had been executed could be distinctly seen." Small lamps marched up columns and spanned archways, like a blinking of fireflies.

These visitors, sardined next to several hundred others, crowded into a room where project plans had been tacked to the wall. Joseph Bazalgette, the chief engineer of the city's sewage enterprise, regaled the gathering with a speech about the building and the broader stakes of investing in the city's waste infrastructure.

There was a lot on the line. The sludge that sluiced through the city's waters was both unsavory and baldly dangerous. London had long been walloped by cholera and other

FIGURE 1 The illuminated underground world of Crossness, pictured in *The Illustrated London News*, April 15, 1865. https://archive.org/details/illustratedlondov46lond/page/348/mode/2up.

diseases we now associate with unsanitary water, and human waste had been front of mind for centuries.[2] An 1189 mayoral proclamation dictated how much space there ought to be between a cesspool and the nearest building, and many more rules rolled out in the following centuries.[3] London's early sewers were only intended to intercept surface water, such as pooling rain.[4] Excretions were sequestered in cesspools, the contents of which were regularly excavated, hauled to the urban outskirts, and sold to farmers, who slathered the mixture on fields.[5] (Officials would keep evangelizing about the potential of sewage-as-fertilizer for centuries: In an 1854 document, civil engineer Thomas Wickstead of the

Metropolitan Commission of Sewers noted that the idea that sewage contained "much valuable fertilizing matter" was "universally admitted," due partly to the presence of "other ingredients from dye works and manufactories, producing refuse of animal and vegetable matter." The "ammoniacal refuse from gas works," he continued, "also adds to its value."[6])

Latrines and cesspools were hardly watertight—they leaked and overflowed, sometimes into the homes of very unhappy neighbors. In October 1660, writes historian Stephen Halliday in *The Great Stink of London: Sir Joseph Bazalgette and the Cleansing of the Victorian Metropolis*, a man named Samuel Pepys (who would go on to become a Member of Parliament) complained to his diary that upon entering his cellar, "I put my feet into a great heap of turds."[7]

Meanwhile, officials skewered the city's existing infrastructure. In an 1842 report about the sanitation landscape, the social reformer Edwin Chadwick concluded that London's sewers amounted to "a vast monument of defective administration, of lavish expenditure and extremely defective execution."[8] The problems he tallied were legion. In some cellars, Chadwick found "nightsoil" heaped up to three feet thick, deep enough to swallow most adults up to the waist.[9] A few years later, an engineer noted other problems, including that one of the sewers had been installed upside down. The egg-shaped tubes were intended to go narrow-end-into-the-earth, so that even slim ribbons of filth would flow; one beneath Chelsea was installed with the broader side down, apparently slowing the festering stream to a dribble.[10]

Pipe problems only worsened when residents started installing toilets. The flushing devices had grown in popularity in Paris during the first few decades of the 19th century, and then made a splash in England at the Great Exhibition of 1851.[11] Visitors to the celebration of industry set up at London's Hyde Park may have returned home jazzed to install a piece of the future. Trouble was, the flushing devices sent more water coursing into the system.[12] At this point, it was no longer illegal to connect household drains to sewers, so the contents often went splashing toward the city's winding, blue ribbons instead of heaping higher at home.

Sewer woes certainly weren't unique to London. Writing about the "intestines" carrying "the very substance of the people" beneath Paris in the first half of the 19th century, Victor Hugo lamented "the wretched vomiting of our sewers into our rivers and the gigantic vomiting of our rivers into the oceans."[13] Boston had been smelly for centuries, too. Soon after sailing from England, 17th-century colonists made unsanctioned sewers by digging rogue brick-and-slate channels near their homes, then connecting them.[14] Like many other sewers, Boston's earliest iterations were gravity-fed, meaning that sewage was urged along by downward slopes. Such systems can sometimes be sluggish, and in a report, the City Board of Health observed that portions of the city were frequently "enveloped in an atmosphere of stench so strong as to arouse the sleeping, terrify the weak, and nauseate and exasperate everybody."[15]

By the mid-19th century, London's young sewers were basically open brooks, discharging right into the waterways, and residents noticed. The Thames and Fleet rivers sometimes seemed to be more waste than water, and cartoonists imagined them harboring terrible, mutant creatures. Artists drew threatening fish with barbed tails and grimacing mouths, or eel-men rising from the murky water, dripping filth. One cartoon referred to the river as "Monster Soup."[16] Speaking of the mucky morass, the famed 19th-century London builder Thomas Cubitt once groused, "the Thames is now made a great cesspool instead of each person having one of his own."[17]

Locals groaned about the stench and beseeched officials to restore some dignity to the Thames. "Where are ye, ye civil engineers?" complained the author of an 1850s column. "Ye can remove mountains, bridge seas and fill rivers . . . can ye not purify the Thames, and so render your own city habitable?"[18]

Before people could picture microorganisms, they could sniff revolting smells, and many subscribed to the miasma theory of disease, which equated rank odors with danger. It was easy to imagine vapors as villains, the stink infiltrating nostrils and fouling bodies from the inside. Many people who caught a whiff of the city's stench worried it would harm them. And they weren't totally off-base: The city's cholera troubles were a product of its poo problem, and the habit of unknowingly consuming feces-flecked drinking water.

Unfortunately for Londoners, things got even grosser as the city grew. In 1855, the scientist Michael Faraday described the water of the Thames as a rather unwelcoming "opaque brown fluid."[19] In the summer of 1858, the funk was inescapable. The sludge-choked waters and fetid riverbanks baked in the dry heat. *Punch* magazine published an illustration of a grim skeleton rowing a wooden boat across the river's surface, passing a dead menagerie with bloated bellies bobbing.[20] The river emitted a stench so awful that the summer came to be known as The Great Stink.

The smell prickled the noses of most everyone in the city—including the politicians sweating in the Houses of

FIGURE 2 In this cartoon from *Punch* magazine, "The Silent Highwayman" (or "Death") rows along the Thames in the midst of the Great Stink. (Public domain, https://commons.wikimedia.org/wiki/File:The_silent_highwayman.jpg).

Parliament along the river's banks, where the stench prodded political will. In an attempt to keep the stink at bay, the window dressings at Parliament were soaked in chloride of lime. It wasn't enough. The leader of the House "was seen fleeing from the chamber, his handkerchief to his nose," Halliday writes.[21] While most residents had no choice but to stick around, politicians debated relocating their own digs elsewhere, farther from the river's reach.

Citywide updates were in order. Bazalgette's team of engineers was devising a new sewer system, but it was a tremendous undertaking. Meanwhile, locals continued to grumble. The month the Crossness facility opened, *The Times* was still describing the Thames with disgust as "a great open sewer, running through the centre of the metropolis, and poisoning the atmosphere with its noisome exhalations."[22] Gesturing from his position in the hall on Crossness's opening day, Bazalgette addressed the ongoing pestilence, and pinned it on poor sewage drainage.[23] To those clustered in the grand chamber, the implication must have been clear: *Look around you. Things are about to improve.*

Crossness is dazzling in size and dizzying in its riot of details, many of which were dreamed up by an architect named Charles Driver who intended to impress.[24] There was nothing humble about the pastiche of Norman, medieval Italian, and Flemish motifs, or the way that the ornate wrought and cast iron were splashed with paint. The *Standard* newspaper mused that Crossness seemed to be conjured by an "enchanter's wand." A few flicks turned the structure into what that reporter called a "perfect shrine of machinery."

The wand was soon waved with even more abandon. The Abbey Mills Pumping Station would be even more lavish. Crossness's younger cousin on the north side of the Thames, Abbey Mills—opened three years later—was (and remains) outfitted with deliciously complicated brackets and spandrels, which help support the roof.[25] The ironwork includes sprays of roses and mounds of leaves, plus clusters of hops and berries. Roundels built into the cast-iron railings bear coats of arms flanked by lilies, and the external domes and now-demolished chimneys were ostentatious enough to earn the facility nicknames including "a mosque in the swamp" and "a cathedral of sewage."[26]

Not all pumping stations of the era were this grand; plenty were comparatively humdrum. Maybe the operations at Crossness and Abbey Mills were deemed glorious enough to be worthy of extravagant flourishes.

Crossness was monumental in several ways. In addition to its massive size—sewage arriving from across the whole of south London was pumped into an enormous covered reservoir that stretched 6.5 acres and could accommodate 25 million gallons—the facility was the centerpiece of workers' lives. Because the plant was remote and the flow of sewage unrelenting, employees lived in on-site cottages and educated their children in a makeshift school in the facility's Valve House.

Crossness impressed onlookers with its steam engines, but also relied on gravity and the tides. When the Thames

FIGURE 3 The Prince Consort engine at Crossness, pictured in 2012. (Nathusius2, licensed under CC by-SA 4.0, https://commons.wikimedia.org/wiki/File:Crossness_Prince_Consort_Beam_Engine_IMG5999-03.jpg).

ebbed, the stored waste would be released from the reservoir through two outfall pipes, "hopefully not returning on the incoming tide," quipped Petra Cox, the learning and outreach manager at the Crossness Engines Trust, which now manages the site. The idea was that the tidal river would pull the pee and poo out to sea and spare London any mess. Instead of eliminating sewage problems, Bazalgette's system shoved them downstream—but for many Londoners, the massive facility was a marvel.

Newspapers had chronicled the construction and debut, keeping residents posted on the progress and operations with sweeping, grandiose images and the play-by-play details present-day readers might expect from coverage of the Olympics. *The Illustrated London News* ran some sketches that sprawled across nearly an entire printed page. In one, a chimney thrusts up into a sky swirling with gray, eddying clouds evoking the waterways that would soon be much less polluted with sewage.[27] There were also careful schematics of the facility's coal vault, engines, and more.

Humans have concocted all sorts of ceremonies to welcome newborns as wiggling little bundles of promise. Londoners feted their nascent sewer. Architecture historian Paul Dobraszczyk argues that both Crossness and Abbey Mills were "key symbolic sites for public awareness" surrounding the city's reborn sewer system.[28] Officials and journalists thought the new modernization effort was well worth celebrating.

That April day at Crossness, the Prince was on hand to turn on the engines, rattling the whole operation into action. Reporters agreed that the system's inauguration warranted the royal touch. "The occasion deserves the honour," *The Times* wrote on April 4. "It represents the accomplishment of a stupendous enterprise conceived in the best interests of social progress."[29]

The Morning Post considered the royal visit to be part of a long, international tradition of taking pride in public works, with "precedents in the care and superintendence which successive Roman emperors bestowed on the *cloaca maxima*"—Rome's "greatest sewer"—and other infrastructure that "facilitated locomotion, improved the health, or added to the comforts of the people."[30] Investing in the sewers was a patriotic gesture.

The prince nudged one handle, then another, and the "stupendous works" began to quake. Cheers erupted from the galleries and men waved their top hats above the crowd. The building rumbled slightly, *The Illustrated London News* reported, "showing that the enormous beams, lifting-rods, and fly-wheels were in operation, and that the sewage which had been confined in the great receptacles underground was being pumped into the reservoirs, thence to be afterwards discharged into the Thames."[31]

Once the machines had been set in motion, some 500 guests decamped for "a capital luncheon."[32] They raised their brimming glasses to toast the Queen and the Prince—and the Prince saluted the future.

"Success to the great and national undertaking, the completion of which we have witnessed this day," he said, according to *The Illustrated London News*. The prince envisioned the sewer as a civic triumph, and a precursor to a different kind of city. "The work will be of material use to London, not so much now, perhaps, as in the future," he continued, "when I hope London will have become one of the healthiest cities in Europe." The future-looking language highlighted ambitions to scale London up to flourish in a modernizing world. The sewer was key to fashioning the metropolis into a bigger, better version of itself.

* * *

The tradition of touting sewers as centerpieces of civic life and pride has deep roots.

When the 18th-century artist Giovanni Battista Piranesi canvassed Rome in search of ruins, he honed in on all sorts of sites of marvelous decay. His multi-volume collection of etchings included detailed renderings of the Corinthian capitals perched atop the columns of the Temple of Juturna, and the plants that had claimed the misidentified, crumbling third-century Temple of Minerva Medica (which was actually devoted to nymphs). Piranesi documented the city's architectural wonders, proof of its history, bustle, and power—and he considered Rome's ancient sewer to be part of that legacy.

A handful of his engravings depict the Cloaca Maxima, the structure that would later be invoked in descriptions of London's expansive new system. (Though, as Roman historian Ann Olga Koloski-Ostrow outlines in *The Archaeology of Sanitation in Roman Italy: Toilets, Sewers, and Water Systems*, this structure wasn't exactly a sewer the way we imagine it today—at least not at first. Though trash and wastewater did sometimes tumble through the structure and into the Tiber River, the system was initially tasked with carrying excess water and dirt off city streets.[33])

In some of Piranesi's etchings, human figures clamber on the structure. At first glance, it's easy to miss them. Their narrow bodies are overwhelmed by the vastness and overshadowed by the maw of the sewer, through which water pours into the Tiber. Piranesi depicts a scene that's a little rough around the edges: Some of the stones on top of the structure were worn down over time into the shape of shrugging shoulders; between them, plants thrust through cracks. But it is grand, enduring, monumental. By including the sewer in his catalog of the city's ancient architectural wonders, Piranesi signaled its importance.

Even much later, sewers have often been central to a city's broader modernization projects. When Georges-Eugène Haussmann, steward of a vast effort to overhaul 19th-century Paris, envisioned tweaks to the medieval version of the city, he pictured graceful boulevards, leafy parks, and an intricate, robust sewer network. He described the pipes as "the internal organs of the great city," and eventually, these organs were opened for tours.[34] By the middle of the century, gawkers could glide through Paris's guts in boats or carts. Guidebooks encouraged visitors, including women in heels and long dresses, to spend an hour or so down there, marveling at the bustling and surprisingly clean operations.[35] "Part of opening the sewers was to say, 'Look at this magnificent achievement,'" Donald Reid, a historian at the University of North Carolina, Chapel Hill, and author of the book *Paris Sewers and Sewermen: Realities and Representations,* once explained to the *Orlando Sentinel.*[36] "It was the Second Empire showing itself off."

Public toilets, too, became a yardstick against which cities could measure themselves and each other. Paris's toilets inspired Londoners of the 1850s, who planned to pull off an international exhibition to rival previous ones on the

Continent. Organizers made the case that sophisticated public toilets were, as architectural historian Barbara Penner has put it, "an index of metropolitan civility." If Parisians had dotted their city with toilet facilities, London had to follow suit—and so the festivities included these facilities, reportedly visited more than 827,800 times.[37]

In other places, sewers became tokens of urban planners' efforts to tame landscapes and extract resources from them—almost-extensions of manifest destiny, expanding to the world beneath our feet. In the mid-1800s, the engineers of Chicago, insistent upon building their city atop soggy wetlands, found that they couldn't install gravity-fed sewers; the land was too close to the water to get the slope right. "So," journalist Dan Egan recently wrote in the *New York Times*, "Chicago's leaders got creative." Instead of inserting sewers under the streets, they plunked them on top, then built new roads above them. As Egan put it, they "effectively hoisted the city out of the swamp."[38] As engineers lifted the city, they also made space for imagining a new future: Midwifed by ingenuity and no small dash of hubris, Chicago would be a vibrant portal to the interior of the country. Egan cites an unnamed 1898 author, who described Chicago as a city that "stands as a stupendous piece of blasphemy against nature." The sewer system was audacious—blasphemous, maybe, but also boldly optimistic.

The ostentatiousness of sewer infrastructure has shifted across space and time—and when sewers and treatment plants became less visibly striking above-ground, some city officials grumbled that their constituents wouldn't be as astonished by the systems as they ought to be. In early-20th-century New England, officials who believed sewers to be engineering wonders worried that the public would take them for granted. As Calvin Hendrick, then Chief Engineer of the Baltimore Sewerage Commission, reflected on the installation-in-progress, in 1909, he apparently fretted that the people using the system would never appreciate it. "The public would be amazed," he reportedly said, "if they could realize what has been accomplished."[39] Roaming the city without the benefit of X-ray vision, they almost certainly wouldn't.

As the 20th century melted into the 21st and many older sewer systems were ripe for upgrades, some officials took the renovations as opportunities to shove their sewers into the spotlight.

Planners at New York City's Newtown Creek facility—which, at 53 acres, is the largest of 14 such stations in the city—did so literally, recruiting a high-profile lighting designer to illuminate infrastructural additions that went up as part of a $5 billion upgrade that nearly doubled the campus's footprint. Solid sludge is now heated up in eight stainless steel "digester eggs," where anaerobic bacteria go to town on it for several weeks. (The final products are biosolids—farmed out as potential fertilizer, for instance, or sometimes shipped to landfills—and methane. A portion runs the boilers, and some will soon be zapped into the nearby community through a partnership with National Grid.) The team enlisted the lighting firm L'Observatoire

International, known for illuminating markedly-less-smelly New York City establishments including the Alice Tully Hall at Lincoln Center, the High Line rail-trail, the swanky Standard Hotel, and portions of the Hudson Yards complex, a shiny capitalist carnival on the city's western edge. The wastewater facility needed to be illuminated for safety, but the team behind the project thought that the glow could also cast it in a metaphorical new light.

In a public statement, the Department of Environmental Protection (DEP) noted that each egg—145 feet tall and weighing two million pounds when empty, or 32 million when stuffed with sludge—would "serve as a landmark" for drivers inching along local expressways or the Kosciuszko Bridge, which shunts travelers from Brooklyn to Queens.[40] The lighting scheme debuted in early June 2008. When the sun set, the lights bathed the hulking ovals in blue. By swathing the eggs in light, a treatment also given to some of the city's more famously architectural skyscrapers, the DEP signaled that the wastewater facility deserved a spot in the urban skyline.

The Newtown upgrade also introduced publicly accessible green space and a visitors' center. Opened in 2010, the center includes educational exhibits, an art installation, and classroom space to cater to the 10,000 visitors—typically students and educators—who tour the center in an average year. The agency developed hands-on curricula to teach visitors about how New York City's water system works and deliver a civics lesson, nudging them to reflect on whether they're cooperating with the system or taxing it. Middle and high-school students physically mimic combined sewers—where wastewater and stormwater merge—by mixing leaves, bottle caps, bits of newspaper, cooking oil, and scraps of plastic bags in a small, water-filled jar, then brainstorming ways to separate and treat the sample. Other activities touch on water conservation, rain gardens and other water-catching infrastructure, and wastewater resource recovery (that is, harvesting usable products from the sewage, including renewable energy). "We want [visitors] to start to digest some of that and understand what the actual steps are, and how we use physical science, biological science, and chemical science [to clean the water]," said Robin Sanchez, Director of Education at the DEP. Reactions vary, Sanchez added: There's "a little bit of surprise and shock—a little 'oohs' and 'ahh'—and sometimes faces of 'eww, that's gross,' but then 'eww, that's gross but it's also awesome.'"

In terms of aesthetic investment, it's not just Newtown. Other NYC facilities are ornamented with city seals, stained glass windows, and more. Brooklyn's Avenue V pump station, recently restored and updated in a $210 million renovation, resembles one of the city's beautiful, Beaux Arts subway terminals.[41] There are tiled mosaics on the walls, and, outside, a scalloped terracotta roof. The building is "a weird little gem that pumps poop," said Alicia West, of the DEP.

When we chatted over Zoom in the spring of 2021, West's background picture was a local frieze that evoked Diego

Rivera's industrial scenes, such as the mural of assembly-line workers that occupies an entire light-flooded room in the Detroit Institute of Arts. There, any gallery goers can appreciate it as they float from Impressionism to ancient Greece and Rome. The image in West's background, on the other hand, is accessible to very few: it's at the Bowery Bay wastewater treatment plant in Queens, glimpsable by employees and passersby who view it through the fence. Sculpted by Cesare Stea as a 1930s Works Progress Administration commission, the frieze shows men lunging, jackhammering, excavating, turning valves. The scene is dramatic, its subjects angular, epic, heroic—the whole thing a bit of artistically scatological propaganda valorizing the dirty work of making the city run.[42]

"Imagine you're a sewer treatment worker," West said. "This would probably make you feel really great." The city's DEP has been explicit about the goal of stoking civic responsibility and making people proud of their town. West traces that ethos back to the construction of the Croton Aqueduct, the massive 19th-century system that ferried water from Westchester south to Manhattan. Both the aqueduct and sewer system were "enormous, unprecedented civic undertakings, without which NYC could never be the city that it is," West said. "I think people had an understanding that, 'That's a big deal, and if we're going to do this, we're going to build these structures that are inherently monumental, then they should really be monumental. They should have gravitas; they should have a civic presence.'"

The Newtown Creek facility has some degree of gravitas just by virtue of its size. When it came to renovations, West said the team figured, "It's so big, you can't hide it, so let's make it look awesome." The design elements also "created a local lore" about the vivid blue eggs, West told me. "You hear silly stories," she said. "Someone once told me that they were bulk onion storage. It gets people thinking about it, even if they don't know what it is." And West thinks that's helpful: "As a city utility, we want people to have an investment in the structures we're building and the infrastructure that serves them. The architecture of the plant serves to pique people's interest, and that's an avenue for talking about how the city works."

The Newtown Creek facility hosts an annual Valentine's Day tour, which migrated to Zoom in 2021 during the COVID-19 pandemic. The digital version was so popular that the first tour maxed out at 1,000 attendees, and the organizers added a second to accommodate the overflow crowd. People tuned in from several states and countries; some were repeat customers who had previously visited in person. Unlike in other years, no one went home with the memory of the sweeping views from the top of the digester eggs or with a Hershey's kiss, a little drop of chocolate redolent of the poop emoji. But they did log off with a much clearer understanding of how the muck snakes beneath them and why they might take pride in the pipes and facilities that make it possible.

Other contemporary cities celebrate the contributions of sewers to civic life in artistic ways. About 23 feet below the streets of the German city of Cologne, a chandelier dangles from the ceiling, flinging soft light into the darkness. The sturdy metal loops and swoops, and each silver arm is crowned with a bulb. The glow washes over archways fashioned from glazed bricks, making the walls seem lit from the inside. The sheen gives the impression of dampness, as if someone has just hosed the walls to rid them of dirt.

Those amber puddles and sense of wetness are totally appropriate for the location. The chandelier illuminates a sewer tunnel. Under the fixture, a narrow, gray river wanders into the darkness. With 15-foot ceilings, this portion of the city's underground network is fairly cavernous. An overflow chamber for wastewater and stormwater, the room was part of the city's 19th-century expansion of its overtaxed sewer system, a warren that struggled to accommodate the needs of a growing city straining the seams of its existing infrastructure. When work on this segment was completed in 1890, officials lauded it with a large festival. Sconces were mounted on the walls, and two chandeliers—predecessors of the contemporary light fixture—were installed from the ceiling. Some sources say that these candlelit ancestors were congratulatory gifts from Kaiser Wilhelm II, while others suggest that city officials installed them to celebrate the emperor, whom they hoped to impress with their innovative labyrinth.

The tunnels were a feat—something to be proud of— and to emphasize the accomplishment, one wall holds an engraved list of the names of architects and the city's mayor, surrounded by an ornamented stone frame that would look at home in a museum or palace. Eventually, the original chandeliers were replaced by an electrified newcomer. More than 130 years later, guided tours bring visitors ambling through the passage. "It smells a little like sewage, but not very strong," explained Stefan Schmitz, who coordinates the tours and gracefully responds to strangers emailing to ask about poop. "If you are in the room for five minutes, you no longer notice the smell." Sometimes, dozens of guests pack into the chamber—which can accommodate 70—for a sewer serenade. Just to the right of the meandering murk, musicians play jazz or classical concerts in the light of the grand fixture, water rushing in the background.

Some communities have installed above-ground monuments to the employees who keep things moving. In the Slovakian city of Bratislava, pedestrians sometimes trip over *Cumil*, a bronze statue emerging from a manhole. Made by the sculptor Viktor Hulík, *Cumil* rests his hard-hatted head on folded hands pressed against the sidewalk, a mooning expression frozen on his face, as if he's recuperating after a long day of trying to defeat a clog.[43] To stop drivers and pedestrians from treading on the statue—which has allegedly been accidentally beheaded—a warning sign was installed nearby. It's a red-and-white triangle, a lot like a "yield" sign, but with an image of *Cumil* in the middle. Beneath the icon is the phrase "Man at work," a nod to the civic work the sculpture celebrates.

The German city of Giessen is also home to a statue of someone with stinky expertise. In the era before indoor plumbing, local workers hauled feces and trash with buckets wedged in narrow gaps between houses. They wielded long, hooked tools to fish the buckets from the crevices and tip them into carts. The statue, installed in 2005, depicts one of the workers ready to schlep his malodorous bounty: He's gripping the tool of his trade and a bucket sits beside him. With rolled-up shirtsleeves and pants stuffed into shin-high boots, he is prepared to wade into a loathsome lagoon.

The Canadian city of Hamilton recently added two 19th-century maintenance hole covers (a gender-neutral term that has replaced "manhole" in several places) to the local registry of objects with cultural heritage value.[44] And London, England, a city plagued by fatbergs—the horrendously hefty hunks of fats, cooking oils, and trash sent swirling down drains and into sewers—installed a decorative cover to commemorate a particular battle in its war on stoppered pipes. In the Whitechapel neighborhood, crews once vanquished a subterranean behemoth that spanned more than 820 feet and weighed some 130 tons. They demolished it over the course of several months, and then shipped portions off to labs and museums for study and gawking. To celebrate the victory over the stubborn blockage, a local utility, Thames Water, installed a cover that reads, "The Whitechapel fatberg was defeated here in 2017." Unlike plaques describing a figure or event that wrinkled the fabric of history, this one hails the often-unsung efforts that keep the city flowing. It highlights the smelly, unseen work as worthy of public recognition, and insists the labor deserves to stop someone mid-stride.

Crossness continues to be a monument to civic achievement, too. More than 155 years after it opened, the facility can still draw a crowd. In a typical year, Crossness opens to visitors around a dozen times. Attendance fluctuates, driven in part by whether the restored engine, dubbed the "Prince Consort," is steaming. The machine no longer pumps sewage, but the Crossness Engines Trust, the custodians of the massive building, occasionally turn it on so that visitors can appreciate the choreography. On days when it's revved up, a thousand or so people might flock to see it. "They want to see it move, and really, I don't blame them," Cox said. Seeing the engine cool, silent, and still, she added, is "like seeing a car and never seeing it drive."

Visitors include "diehard engineers who will go anywhere to see a steam engine in action," she said, plus architecture students, urban planners, environmental groups, nursing and medical school students, and families. "We're coming from a common ground with everybody," added Cox, who came aboard to help interpret the site for a general audience. "Everybody's crafting the problem"—meaning poop— "and this is the Victorian solution." Visitors encounter an exhibition about the Great Stink, which tracks the arc of sanitation infrastructure.

That arc continues into the 21st century. Bazalgette's monumental project didn't fix the conundrum of what to do with feces. Facilities are still inundated with more than they can handle; waterways are still sludgy with overflow. The problem "never gets solved," Cox said. "It's needing to get updated, and continuing all the time." Cox's team worked on a digital "toilet timeline," winding from the contraption's early days to current shortcomings around the world, as well as possible improvements.[45] Cox thinks that many people have embraced the global need for clean drinking water, but don't give as much thought to sanitation as it relates to sewers. "Sanitation is so unglamorous," she told me. "It's the poor cousin of the lovely water wells and clean drinking water."

Cox thinks that needs to change: The attitude that sewage is a topic to be discussed in embarrassed whispers, if at all, "is kind of strange to me, considering that we're all in this together," Cox added. She's a promoter of, as she put it, "responsible citizenship," which includes using water in ways that don't heap stress onto the sewer system. And while the coronavirus pandemic temporarily paused public tours, Cox began working on a virtual version, to launch in spring 2022. No one has to float down the Thames to learn about the enormous engines. They can study London's relationship with wastewater from the comfort of home—the place where the messy business begins, and the place where stewardship should start. No one can build a Crossness in their house, and that's fine. Instead of constructing a temple to infrastructure, we can be allies of the systems that are already in place.

Notes

1   "The Prince of Wales at the Metropolitan Drainage Works," *The Illustrated London News*, April 15, 1865, 342.

2   For more on this, see: Steven Johnson, *The Ghost Map: The Story of London's Most Terrifying Epidemic—and How It Changed Science, Cities, and the Modern World* (New York: Riverhead Books, 2007).

3   Stephen Halliday, *The Great Stink of London: Sir Joseph Bazalgette and the Cleansing of the Victorian Metropolis* (Thrupp: Sutton Publishing, 2001), 30.

4   Ibid., 28.

5   Ibid., 31.

6   Thomas Wickstead, *Report Upon the Most Advantageous Mode of Dealing With the Sewage Matter of the Metropolis With a View to the Preparation of Sewage Manure* (London: James Truscott, 1854), 6, London Metropolitan Archives, City of London COL/TSD/EG/05/01/024, from the Corporation of London collection.

7   Halliday, 32.

8   Ibid., 39.

9   Ibid., 40.

10 Ibid.

11 Ibid., 43.

12 Ibid., 43–46.

13 Victor Hugo, *Les Misérables*, Part 5, Book II, trans. Sir Lascelles Wraxall (Boston: Little, Brown, and Company, 1887), accessed via Project Gutenberg.

14 Miriam Wasser, "Boston's Stormwater Problems Are Older Than The City Itself," WBUR, June 17, 2021, https://www.wbur.org/news/2021/06/17/boston-stormwater-sewer-history.

15 Elliot C. Clarke, *Main Drainage Works of the City of Boston (Massachusetts, U.S.A.)* (Boston: Rockwell and Churchill, City Printers, 1885), 14.

16 William Heath, *A woman dropping her porcelain tea-cup in horror upon discovering the monstrous contents of a magnified drop of Thames water; revealing the impurity of London drinking water*, 1828, colored etching, Wellcome Collection.

17 Ferdinand Mount, "Lost in a Time of Cholera," *Wall Street Journal*, October 21, 2006, https://www.wsj.com/articles/SB116138138448699373.

18 Halliday, 26.

19 Jonathan Schneer, *The Thames* (New Haven: Yale University Press, 2005), 147.

20 "The silent highwayman," *Punch Magazine*, July 10, 1858, 137.

21 Halliday, 17–18.

22 *The Times*, April 4, 1865, 11.

23 "The Prince of Wales at the Metropolitan Drainage Works," *The Illustrated London News*, April 15, 1865, 342.

24 Paul Dobraszczyk, "Architecture, Ornament and Excrement: The Crossness and Abbey Mills Pumping Stations," *The Journal of Architecture* 12, no. 4 (October 2007): 353–65, https://doi.org/10.1080/13602360701614631.

25 Paul Dobraszczyk, "Historicizing Iron: Charles Driver and the Abbey Mills Pumping Station (1865–68)," *Architectural History* 49 (2006): 223–56, https://doi.org/10.1017/S0066622X0000277X.

26 Paul Dobraszczyk, *London's Sewers* (Oxford: Shire Publications, 2014), 48.

27 "The Metropolitan Main Drainage—General View of the Southern Outfall Works at Crossness," *The Illustrated London News*, April 4, 1865, 325.

28 Dobraszczyk 2007.

29 *The Times*, April 4, 1865, 11.

30 *The Morning Post*, April 14, 1865.

31 "The Prince of Wales at the Metropolitan Drainage Works," *The Illustrated London News*, April 15, 1865, 342.

32 "Summary of This Morning's News," *Pall Mall Gazette*, April 5, 1865, 5.

33 Ann Olga Koloski-Ostrow, *The Archaeology of Sanitation in Roman Italy: Toilets, Sewers, and Water Systems* (Chapel Hill: The University of North Carolina Press, 2015): 63–66.

34 Stephen Halliday, "On (and In) The Sewers (and Sewage) That Transformed Paris," *Literary Hub*, December 2, 2019, https://lithub.com/on-and-in-the-sewers-and-sewage-that-transformed-paris/.

35 Donald Reid, *Paris Sewers and Sewermen: Realities and Representations* (Cambridge: Harvard University Press, 1991), 41.

36 Loraine O'Connell, "The Paris Sewer System: A Breathtaking Attraction," *Orlando Sentinel*, January 7, 1996, https://www

.orlandosentinel.com/news/os-xpm-1996-01-07-9601020535
-story.html.

37 Barbara Penner, "The First Public Toilet?: Rose Street, Soho,"
Victorian Review 39, no. 1 (Spring 2013): 26–30, https://www
.jstor.org/stable/24496993.

38 Dan Egan, "A Battle Between a Great City and a Great Lake,"
New York Times, July 7, 2021, https://www.nytimes.com/
interactive/2021/07/07/climate/chicago-river-lake-michigan
.html.

39 Alicia Puglionesi, "The Manmade Marvel of the Baltimore
Sewers," *Atlas Obscura*, January 6, 2016, https://www
.atlasobscura.com/articles/the-manmade-marvel-of-the
-baltimore-sewers.

40 New York City Department of Environmental Protection,
"DEP Celebrates Lighting of Newtown Creek 'Digester Eggs'
Landmark," news release, June 3, 2008, https://www1.nyc.gov/
html/dep/html/press_releases/08-14pr.shtml.

41 New York City Department of Environmental Protection,
"$210 Million Upgrade of Historic Beaux-Arts Style Pump
Station in South Brooklyn Substantially Improves the Health
of Coney Island Creek," news release, March 24, 2015, https://
www1.nyc.gov/html/dep/html/press_releases/15-016pr
.shtml#.YQ8tMtNKhHR.

42 Paul Wood et al., *Modernism in Dispute: Art Since the Forties*
(New Haven: Yale University Press, 1993), 13.

43 "Cumil the Sewer Worker," *Atlas Obscura*, https://www
.atlasobscura.com/places/cumil-the-sewer-worker, accessed
January 6, 2022.

44 Matthew Van Dongen, "Hamilton Bestows Protection on Pre-
Confederation Era Manholes," *Toronto Star*, March 16, 2018,
https://www.thestar.com/news/gta/2018/03/16/hamilton

-bestows-protection-on-pre-confederation-era-manholes
.html.

45 For more on this, see: Chelsea Wald, *Pipe Dreams: The Urgent Global Quest to Transform the Toilet* (New York: Avid Reader Press, 2021).

2 WIPES & PIPES

How tiny cloths can cause subterranean chaos.

John French's entanglement with wet wipes started out as a joke. A couple decades ago, he and a buddy named Jim turned them into a recurring gag. The pals were space nerds, and French still is; he's now the program assistant at Michigan State University's Abrams Planetarium in East Lansing. Back then, whenever he and his friends saw a planetarium show that highlighted how amazing it is that we all live on a tiny, marbled speck in a solar system—just a sliver of a galaxy, which itself has billions of galactic neighbors holding untold planets—they would mime the sensation of being misty eyed, choked up, *verklempt*.

"We'd say, 'I need a moist towelette,'" French recalled recently, grinning and fanning his face with one of his hands. The joke was that a towelette would be calming and refreshing, something nice to dab around your eyes.

French, who sometimes accompanies planetarium shows on his ukulele, is 60, but still possesses the playfulness of a middle school prankster. When I met him on a warm December day in 2019, he had capped his silver-and-gray shag with a red knit hat and pulled on a long-sleeved waffle shirt and a purple-and-blue flannel. He's a guy who appreciates whimsy: His office, a jumble of tchotchkes, contains a little figurine carved from a hunk of desiccated Spam.

At the Planetarium, French is a custodian of history both natural and supremely synthetic. He narrates shows about the wonders of the night sky, murmuring through a microphone in a soothing, late-night-public-radio voice. He flicks on lights that illuminate the chunk of the pitted, football-sized meteorite that once thwacked down in Edmore, Michigan, and the fish tank-style vitrine where visitors can use a remote control to steer a replica Mars rover over a swath of sand. He leads guests through the black light gallery, where they can look out over craggy extraterrestrial landscapes that glow in theatrical, not-exactly-accurate hues of bright pink and green.

Wet wipes don't have much to do with the stars or planets, but French's office is full of them. On two overflowing filing cabinets outfitted with black velour and little lights, he displays his collection of several hundred packets. Inside the planetarium, French moonlights as the curator of what is probably the world's only Moist Towelette Museum. The quirky collection, open by appointment, is proof that wipes are ubiquitous across the world—often marketed as throwaway items, designed to meet a pleasantly watery end.

The museum is funny, charming, benign. In pipes, wipes are often none of the above. Because the cloths are a lot smaller and flimsier than the tubes they enter, it might seem as though the thin, scrunchable sheets ought to be insignificant specks in the underground cosmos. But it doesn't usually work out that way.

Sewers are designed to intercept urine, feces, toilet paper, and vomit, but all sorts of alien things, from wet wipes to alligators, wind up in the pipes. Creatures—real, imagined, or occupying some murky place between the two—have been associated with sewers for millennia. The ancient Roman writer Aelian, for one, described an octopus of "monstrous bulk" that, preferring to scrounge for lunch on land, swam up through sewers, heaved its body across the floor, and helped itself to the pickled food that merchants stored in earthenware jars.[1] Later, in 19th-century London, a group of "toshers"—scavengers who canvassed the sewers or the stinking shore of the Thames for metal or other things to rinse and sell—reportedly told the journalist Henry Mayhew that a sow had birthed a litter in the sewer. (The piglets supposedly supped "on the offal and garbage washed into it."[2]) A similar story surfaced in an 1859 article in the *Daily Telegraph*, which reported that some residents were convinced that a "monstrous breed of black swine" was thriving in a sewer near Hampstead.[3]

Tales about subterranean species continued into the 20th century.[4] And while much of the critter-lore is exaggerated or baseless—though some pigs may have once escaped their fates in English meat markets, it's exceedingly unlikely that a population really ran amok beneath London—other anecdotes are grounded in fact. Modern-day whip spiders—arachnids with long, spindly, sharply angled legs—have been known to thrive in Jerusalem's sewers.[5] Rats are longtime residents of wet tunnels, and so are alligators. In February 1935, two kids dumping snow into an East Harlem sewer spotted an eight-foot gator, likely a pet turned loose by its owners. The kids "lassoed" the animal with a clothesline and lugged it up to the street. They eventually mauled it with their shovels, and the *Daily News* published a picture of them holding their bounty.[6] Sightings have continued for decades, and the creatures sometimes surface on the street. "Alligators survived the last Ice Age, so I'm really not in a position to say they can't survive the sewer—they've clearly survived a lot worse," Thomas Hynes, author of *Wild City: A Brief History of New York City in 40 Animals*, once explained.[7] There's plenty for them to eat, and darkness is no bother, since they don't rely on their eyesight to hunt. (An honest real estate agent leveling with a crocodilian about the drawbacks of this potential housing would caution about limited nesting materials and slim pickings for mates.)

Sewers may swarm with life, but creatures don't snarl sewer operations the way trash does: Debris has dogged pipes and treatment plants for centuries. In the 1880s, for instance, a cage called a "filth-hoist" at Boston's sewage treatment works intercepted "rags, paper, corks, bottles, half lemons, lumps of fat, dead animals, pieces of wood, children's toys, pocket-books, and such-like miscellaneous articles," according to a city report.[8] Around the world, utility companies' technicians have more recently fished out everything from lingerie sets and kitty litter to cutlery, phones, toys, dentures, and waterlogged Yorkshire puddings, which looked like soggy croissants garnished with a dollop of dog puke.[9]

Any clog is worth grumbling about, but utility workers save special complaints for wet wipes. A spokesperson for the English utility company Anglian Water has estimated that the moist scraps are at least partly to blame for about 80 percent of the 40,000 blockages the agency fights each year; in some other places, they're the culprits in nearly every backed-up pipe.[10]

It's fitting that French's continent-spanning collection is vast: In 2018, *The Guardian* reported that humankind uses around 450 billion wet wipes each year, or roughly 14,000 each second.[11] Just a few bucks a pack, wet wipes are cheap for consumers, and a massively lucrative industry for the companies that produce them. They're popular among adults and their babies, and cleanse everything from mascara-caked eyelashes to kitchen counters and humans' bottoms. Sewers are teeming with them because humans insist on flushing them—and those cloths can cause subterranean chaos.

* * *

When first freed from its little plastic package, a wet wipe is pristine. The cloth is damp, white, and designed to get dirty in the name of cleanliness.

Though they've seen an uptick in popularity in recent years, wet wipes were invented decades ago. I'm not sure who first hit upon the idea, but the concept dates back more than 70 years. In January 1948, a man named Fritz Freuthal filed a patent in the United Kingdom for an invention he described as "improvements in and relating to cleaning or sanitary wiping material."[12] While he doesn't string the two words together in that now-infamous order, the gist of his invention sounds a lot like wet wipes.

The idea went like this: Saturate a piece of "highly absorbent fibrous material," such as a bit of cellulose wadding, with a disinfectant and a hygroscopic agent, like glycerin. (A hygroscopic agent is one that absorbs and retains moisture; it's what keeps the wet wipe wet.) Reinforce the cloth with something strong and flexible that can withstand the moisture—muslin, maybe, or rubber or waxed paper. Souse the wipe with a pleasant fragrance, courtesy of detergent or perfume, and consider sandwiching the wet cloth between slips of metal foil or waxed paper to keep things from getting too sopping. Tuck it into an envelope, and voila—a wipe that's damp, sanitary, and simple.

The concept took off across industries. Hospitals snapped them up to disinfect surfaces and wipe down medical instruments. (Years before the COVID-19 pandemic inspired many places to adopt even-more-fastidious standards for cleaning, University College Hospital, in London, tore through 90,000 packs of wet wipes annually.) Wipes were also embraced across the food world, especially in restaurants serving stuff slathered in sticky sauces. In the 1950s and 60s, Arthur Julius, who had a background in the cosmetics industry, tinkered with an invention that he dubbed the Wet-Nap. He peddled his invention to Kentucky Fried Chicken (now KFC), and the restaurants were soon sending customers home with wipes, perfect for removing the glistening grease from their fingers after polishing off a bucket of wings. Over the past 25 years, the chain has handed out nearly a billion wipes.[13]

It soon seemed that nearly every field had a need for the cloths. French's collection includes Shout wipes to blot out stains, towelettes for rubbing down dentures without popping them from your mouth, a slightly wrinkled pack from the Hard Rock Cafe in Beijing, and one with pink text, a rose, and a "baby powder scent" promising to impart "confident freshness." There are packets from Lufthansa, British Airways, and Iberia Regional Air, plus a shamrock-spangled version from Aer Lingus. French also has a packet from the old Trump Castle hotel and casino in Atlantic City, reading, in swoopy cursive, "The Trump Card." He has one of those billion wipes from KFC, too, with Colonel Sanders's half-cracked grin.

Wet wipes entered the bathroom partly because advertising campaigns and jingles suggested people's heinies were not getting sufficiently clean with toilet paper alone.

Manufacturers and retailers continue to emphasize the idea that damp cloths are key to cleanliness. In 2019, my dad texted me a photo from a superstore in suburban Michigan. On a grocery-shopping outing, he had noticed that the retailer was selling wet wipes in the toilet paper aisle, in a display case reading, "Flushable wipes + toilet paper = cleaner & fresher." This supposed need for wet wipes is sometimes echoed by medical professionals in articles in women's magazines including *Self*, which once described moist wipes as a way to avoid chafing sensitive nether region skin and lower the risk of accidentally swiping bits of feces toward the urethral opening, where the flecks could potentially cause urinary tract infections.[14]

In America, wet wipes—like their drier counterpart, toilet paper—are more popular butt cleansers than bidets, which haven't quite caught on despite being ubiquitous elsewhere. (Instead, Americans purchase 20 percent of the toilet paper sold around the globe, Jackie Flynn Mogensen reported for *Mother Jones*, despite making up "just 4 percent of the world's butts."[15])

Bidets have a lot going for them. For one, they're easy to use: Many modern versions pop onto existing toilets, attach beneath the seat, draw cool water from the same supply that fills the tank, and have a nozzle that shoots water buttward with the turn of a knob. They're also eco-friendly. The spritzers use around two cups of water per sitting, which can be less than what goes into manufacturing a handful of sheets on a roll of toilet paper, though that mileage varies based on how plush the TP is and whether it was produced from, for instance, recycled materials versus virgin wood from clear-cut forests.[16]

Over the past few years, bidets have earned some nods from high-profile entertainers, filters through which American trends often trickle. Back in 2014, actor Kumail Nanjiani told late-night host Conan O'Brien that his "life will be divided into two sections: Before I ever used a bidet, and the Age of Enlightenment."[17] One bidet manufacturer registered a measurable leap in sales after comedian Michael Che joked, on *Saturday Night Live*, that the contraption was "glorious" and moved him to tears. Still, a survey conducted by the bathroom appliance company Kohler in 2016 found that more than half of American consumers were squeamish about the appliances.[18]

Writing for *The Atlantic* in 2018, Maria Teresa Hart recounted how Americans' aversion to the bidet might trace back to wartime indiscretions and be rooted in misogyny.[19] Soldiers encountered bidets while frequenting European bordellos during World War II and came to associate them with prostitution. The device had also long been linked with other taboos coded as female: Before there was an industry to supply pads, tampons, and other products to absorb menstrual blood, smears could be splashed away by straddling a bidet. It was also associated with contraception: "In the United States and Britain, when various forms of douching were thought of as a pregnancy preventive, bidets were considered a form of birth control" thought to knock sperm off course, Hart wrote. She argued that the bidet's chances in America were stymied by its perceived relationship to all of these "feminine failings."

But there has been incremental interest. Back in 2019, an industry newsletter forecast that US sales would rise by 15 percent annually through 2021.[20] Then came COVID-19. Enthusiasm for bidets surged like a blast of water.

Sales spiked as people prepared for an unknown stretch of quarantine and social distancing. The manufacturer Tushy—which hawks bidets via quippy, crisp, Millennial-friendly marketing—saw sales begin to rise during the first week in March 2020. "Then two days later they were triple what they usually are, and then suddenly it was 10 times what normal sales are," and soon crossed the million-bucks-a-day mark, CEO Jason Ojalvo told *The Guardian*.[21] That same month, a spokesperson for the home goods company Brondell told *Business Insider* that the company was selling a bidet through Amazon every two minutes.[22] As the pandemic wore on and those Americans with the privilege of hunkering down in their homes increasingly opted to overhaul their spaces, some revamped their bathrooms.[23] By November 2020, according to one industry report, bidets were the leading additions to these renovated chambers.[24]

Optimists might expect that this uptick in bidets would lead to a sharp decrease in sewer clogs; cleansing with streams of tepid water would mean fewer additions of paper products to plug up the pipes. That's not what happened. Instead, demand for wet wipes revved up, too. As the pandemic spread, cloths sold like perfumed, microbe-busting hotcakes. Sales surged by 75 percent in the twelve months ending in January 2021, *Bloomberg* reported. (Cynthia Finley, Director of Regulatory Affairs at the National Association of Clean Water Agencies, or NACWA, told me that shoppers were likely spurred by shelves emptied of toilet paper, coupled with the impulse to disinfect everything in sight.) As people clamored to scrub door handles, ATM pads, and their own bodies, those wipes found their way into toilets and pipes.[25]

As a result, the pandemic era has been a bonanza for the clogs wipes can cause. In Des Moines, Iowa, backups climbed 50 percent during the pandemic, *Bloomberg* noted, and officials pinned the blame on wipes.[26] Between April and June 2020, clogs at pump stations in Charleston, South Carolina, leaped eightfold, and a pump station in Clinton Township, northeast of Detroit, intercepted four times more wipes each week than in previous years.[27] In the Maryland suburbs, crews from WSSC Water freed 700 tons of knotted wipes in 2020—100 tons more than the prior year.[28] As communities wrestled with public health crises above-ground, sewers were struggling, too.

*   *   *

For decades, inventors and manufacturers have promised that wipes were fine to flush. The trick, they claimed in patents, was to find a tensile strength that would keep the wipe intact during storage and use while also encouraging the cloth to fall to pieces when it hit the toilet bowl and was agitated by the swirl of the flush.

It was hard to find a material that nailed both, because the binding agents that kept the wipes together in the package often worked too well, and for too long. Some experts even suggested that someone planning to flush a wipe ought to splash some acidic or alkaline substance in the toilet bowl before they dropped the wipe in, just to give the disintegration process a head start or some sharper teeth.[29]

It quickly became clear that "flushable" wipes didn't work quite the way the inventors promised they would. Over several decades, subsequent patents claimed to be improving upon earlier duds. The promise of wet wipes that fully disintegrate and disappear hasn't universally come to pass. But we've kept on flushing them anyway.

In the process of cleansing their rears, wet wipe users sometimes wind up polluting the planet. Around the world, sewers still often discharge into waterways, spitting untreated sewage and stormwater into lakes, rivers, streams, even oceans. Many wipes travel through sewer pipes fully intact, and tumble out with the rest of the waste. They don't immediately disintegrate in those bodies of water, either— instead, they reshape ecosystems.

In London, for instance, tens of millions of tons of rainwater and sewage slop into the River Thames each year when downpours stress the pipes, and those gushes include a lot of wipes. The cloths sometimes become geologic features—new, synthetic elements of the landscape that modify the shoreline and mangle the natural order of things.

The Thames meanders through the city in a squiggle, and because jutting land slows the flow, wipes tend to cluster where its route bends. There are easily thousands—maybe even millions—of wipes lodged near the water. On an embankment near Hammersmith, low tide reveals a shoreline swollen with them, a facsimile of a riverbank. At a shoreline cleanup along the Thames in 2018, volunteers canvassing a 1,250-square-foot patch of ground pulled up close to 5,500 wipes. The following year, in just two hours, volunteers working along a small portion of the river collected an estimated 23,000 wipes—enough to fill 473 trash bags.[30] The problem has continued to worsen. Over two days in 2021, volunteers yanked 27,000 muddy wipes from a 655-foot stretch between Battersea Bridge and Albert Bridge. Bathymetric surveys have revealed that, in some places, the mound of mucky wipes and other sediment-embedded trash stands more than three feet high and stretches more than 160 feet wide.[31]

One ecologist told the *Evening Standard* that, from a distance, the wipes look like seaweed, quivering and dancing in the river's waves.[32] Up close, it's a big mess: The pileup fouls boat propellers and puzzles local birds, which try to perch on the squishy heaps. The feathered fliers reside in both the natural world and the synthetic one we have overlaid atop it. The soggy land is, to borrow from the creepy and poetic phrasing of the nature group Thames21, a "Great Wet Wipe Reef" and a "Frankenstein foreshore"—a patchwork of natural and wildly artificial elements that borders on the freakish.[33]

In other cases, flushed wipes don't go anywhere at all—and that's a big problem, too, in both sewers and septic systems. Wipes' persistent sluggishness sometimes takes them from the toilet to the courtroom.

In the summer of 2013, a dad in Ohio started noticing that something was off with his pipes. The tub was slow to drain; the toilet wouldn't flush properly. That fall, he hired a plumber who quickly determined that the problem was wipes—a whole tangled slew of them.

The dad had been potty-training his young daughter, and mess comes with the territory; kids learning to poo in the loo miss a lot at first. Figuring that wipes would make clean-up easier, the dad stocked up on a house brand from a Target in the town of Boardman. Since the package promised that the wipes were flushable, the dad sent them down the drain. They didn't make it very far. The plumber found that the wipes had smushed together in the pipes and septic system, causing everything to back up. The plumber flushed the system—at a cost of $210—and cautioned the dad that he might need to pony up $20,000 for a new septic set-up if the wipes had permanently damaged the existing one. The full extent of the snarl wasn't yet clear.

Wet-wipe industry lobbying groups such as the Association of the Nonwoven Fabrics Industry, also known as INDA, have argued that wipes aren't the problem—people are, when they flush whatever they want to. Several of the packages in the Moist Towelette Museum indicate that used wipes should be thrown in the trash; a few have little icons of a toilet with a line struck through it, indicating that they don't belong there. (Some researchers argue that the symbols are maddeningly unstandardized and tricky for consumers to parse.) INDA representatives have insisted that the wipes that cause the most trouble are the ones people aren't supposed to flush, but do.

But the Ohio dad thought he was following instructions. Frustrated by the flushable wipes' lack of flushability, he became the lead plaintiff in a class-action suit filed in the U.S. District Court of the Northern District of Ohio on behalf of any consumers who purchased the wipes, complaining that that the "sewer and septic safe" claims were "false" and "misleading."[34] The package promised that the wipes would break down after flushing, but, noted a blurb in the *National Law Review*, "the lawsuit alleges that precisely the opposite is true: that the wipes do not disperse after flushing and instead result in clogged sewer lines and septic systems, causing sewage backups and even flooding."[35]

As the dad in Ohio discovered, wipes that are advertised as sailing smoothly through the pipes don't necessarily fall to pieces the way customers might imagine. But determining whether something is "flushable" is trickier than it might seem. It's not as simple as looking to see whether it slides down the hole in the toilet bowl and disappears from view. A few years ago, researchers in Michigan evaluated six different varieties—a baby wipe, hand wipe, makeup wipe, and more—to see whether they would break down as promised when shaken in beakers on a quaking table (one way that scientists attempt to emulate the movement of wastewater). The researchers waited a day, then two. They waited a week, then three. With the help of the local utility company, they swapped the tap water in the beakers for wastewater, just in case the sewer environment contained bacteria that were key to disintegrating the wipes. Still, nearly none of the wipes broke down.

To suss out whether any really can sluice through the toilet and eventually disintegrate in the pipes or in the sewer, a handful of organizations around the world subject wipes to a battery of tests. (Some experts say these assessments—which involve heating, drying, shaking, and generally pummeling the wipes, over and over again—aren't perfect proxies for conditions in the sewer, which are likely gentler.) In 2019, the trade association Water UK laid out some ground rules for how many flushes it could reasonably take a "flushable" wipe to clear the toilet bowl (two), how far down the drain line a wipe should be able to trek within five flushes (20 meters, or around 66 feet), and how small its remnants should be as it disintegrates in the sewer (mostly small enough to slip through a sieve with holes measuring just 5.6 millimeters across).

The tests were pass/fail—and even when they were repeated multiple times, nearly every wet wipe product on the market floundered. Ones that did pass earned a "Fine to Flush" insignia to splash across their packaging. The first passing wipe was a brand called Natracare, made from paper tissue doused with ingredients including witch hazel and aloe vera. The wipes are plastic-free, and their packaging features a large badge reading "tested for toilet disposal," plus a peaceful whorl of blue-green turtles and schooling fish whose lives would apparently not be ruined if you flushed the wipe down the drain. Since then, others have received this imprimatur handed out by the Water Research Centre, including wipes made for Aldi, Sainsbury's, Tesco, and Waitrose.

Like the U.K., North America also hammers wipes aboveground to see what might happen below. Some utilities have tested wipes against standards outlined by the International Water Services Flushability Group (IWSFG), made up of wastewater companies and clean-water organizations. These tests involve flushing a wipe down a control toilet into a fake drain line, then retrieving it, swirling it in a bucket, and then jiggling it in a plastic contraption known as a "slosh box."[36] (This isn't quite what happens in the sewer, either; sewer pipes typically don't produce much sloshing.) This box is wagged around, and then the contents are stirred, dumped into a pitcher, and poured through a sieve with 25-millimeter holes, which is rinsed with a shower head. The little bits that still cling to the sieve are then scooped up by hand or with tweezers, then dried and weighed. Technicians calculate how completely a wipe disintegrated by comparing the weight of those undissolved bits to the weight of the original, unthrashed sample. For a wipe to pass, at least 80 percent of the bits must squeak through the holes. Individual brands often perform their own tests—sometimes with different protocols and less-stringent standards. (For instance, in 2018, a wipe was considered "flushable" by INDA's definition if 60 percent of it made it through a sieve.[37]) Flushability, it seems, is sometimes in the eye of the flusher.

Expensive clogs are a long-term hassle and bipartisan concern, menacing places that are urban and rural, beachfront or landlocked, conservative or liberal or purple. In 2020, NACWA estimated that wet wipes cost water utilities across the United States an extra $441 million a year. The hassles are most pronounced on the coasts and dense, industrial pockets of the Midwest, but NACWA estimated cost increases pretty much everywhere, and pointed out that those trickled down to consumers.[38] (To compensate for pipe-cloggers, customers in Illinois, for instance, were reportedly getting hit with an extra $25 in costs each year.) In an era of deep chasms of political division, dismay over wet wipes bridges the aisle: Republicans and Democrats alike have thrown their support behind bills around wet-wipe labeling.

Because "flushability" is murky, some states and cities have stepped in to set out some clear rules. In 2020, the state of Washington became the first in the US to pass a law requiring manufacturers to place a "do not flush" label on potential pipe-pluggers.[39] Oregon, California, and Illinois followed suit, and bills have also been introduced in Massachusetts, Minnesota, and Maine, Finley told me. In London, Fleur Anderson, a Member of Parliament, recently proposed a bill that would ban manufacturing and selling wet wipes containing plastic. Thames21 is buoying the effort.

In spring 2020, when wastewater workers in Michigan's Macomb County were fishing about 4,000 pounds from pump facilities each week, public works commissioner Candice Miller announced a lawsuit aiming to compel wet wipe companies to change their labels and indicate that their products shouldn't be flushed. "Use 'em—just don't flush them down the toilet," Miller said in a recorded PSA.[40] "That's all I'm trying to do here, change the packaging." (Wipes in landfills or incinerators create environmental burdens, too, but not ones shouldered by water utilities.)

Other states have gone after specific companies. In April 2021, South Carolina's Charleston Water System negotiated a settlement with Kimberly-Clark Corporation following a 2018 fatberg so gnarly that it could only be loosed by a crew of divers lowered into the darkness on a metal cage.[41] Kitted out in steel-toe boots, hands stuffed into three pairs of gloves, the workers felt their way through Stygian conditions and dismantled a fatberg so shaggy with waterlogged wipes that, when the clog was towed to the surface, it looked like a monstrous, gray-black piñata pounded by rain.[42] The lawsuit says that to be appropriately deemed "flushable," a wipe "needs to be able to disintegrate into smaller pieces rapidly enough to pass through sewer systems and be appropriate for treatment" (much like toilet paper, which begins to fall apart upon meeting water).[43] As part of the settlement, Kimberly-Clark Corporation vowed to bring their "flushable" wipes in line with the IWSFG's standards by 2022.[44] In tests of seven brands in January and February 2021, the Charleston team found that Cottonelle wipes were the only ones currently safe to send down the drain.[45] After the slosh box test, the Cottonelle contender was tattered in small, jagged pieces evoking pulsing, translucent jellyfish; in the sieve stage, nearly nothing stayed behind.

In the summer of 2021, congresspeople from California and Michigan introduced a bipartisan bill to create a national standard for package labeling, hoping to compel manufacturers of non-flushable wipes to slap a "DO NOT FLUSH" warning across their products. (Called the Wastewater Infrastructure Pollution Prevention and Environmental Safety Act, it is charmingly, if somewhat confusingly, abbreviated "WIPPES.") It hasn't yet gone up for a vote, but NACWA "does see support for bills like this growing," Finley said. She figures residents might rally behind the proposals as they hear more about the nuisance wipes can cause for utilities and the people who keep the wastewater moving along.

Many utility companies are skeptical of flushability claims, and several cities have launched public awareness campaigns to discourage residents from flushing any wipes at all. A water company in South Australia made a radio-friendly earworm. The 17-second folksy ditty beseeched residents to only flush the "three pees"—paper, pee, and poop—and to toss everything else in the trash.[46] New York City launched a "trash it, don't flush it" campaign asking locals to toss wipes "even if they're 'flushable.'" Fatberg-busting PSAs once loomed over residents skedaddling through subway stations, a reminder of shenanigans playing out in other subterranean tunnels nearby.

*　*　*

Toward the end of my time with John French—after a ukulele serenade and tour of twinkling wonders beneath the planetarium's dome—I asked him what he thought of wet wipes' status in the sewers. As the caretaker of the collection, how did he feel about the havoc his moist towelettes could cause? "I haven't really given that a whole lot of thought," he said, trailing off. "I do like the environment."

He'd seen documentaries about fatbergs, he continued, and he laughed, as if to say, *yikes*. He wanted me to know that he doesn't flush the cloths. "I don't want to get too personal, but I do not use the wipes at home in the bathroom," he added. He used to, but stopped after learning that the wipes don't break down the way he once assumed they did. He's had enough sewer problems as it is, including tree roots encroaching on the pipes. (For years, he had someone come by to fend off the wandering roots; he has since replaced the sewer line and said goodbye to the offending silver maple.) His wipe collection is for show, not for flushing.

In an era when so much is changing in forests and in oceans, in lakes and rivers and on the land that water laps against, it strikes me as ethically murky to willfully worsen the situation in the name of a cleaner butt, especially when there are viable, sustainable alternatives. Some require money and foresight; others call for nearly none at all. Renters could install bidets on existing toilets, and homeowners could incorporate them into future renovations; everyone could stop flushing anything beyond toilet paper and bodily excretions.

When we send wipes down into the sewers, we can't know where they'll end up, or how, exactly, they might snarl infrastructure or the natural world. But wipes are already mounding up in rivers; our current behaviors are already reshaping the planet. Even if you're always separated from pipes by a cement or asphalt barrier, the environmental fallout is all around us—partly because so many wipes live underground or on shores, not just in John French's lovingly curated collection. If they had all stayed there—goofy packages gathering dust in an overstuffed office—pipes everywhere would be much better off.

Notes

1   Aelian, *On the Characteristics of Animals, Volume III, Books 12-17 (Loeb Classical Library No. 449)*, trans. A. F. Scholfield (Cambridge: Harvard University Press, 1959), book 13. This story also appears in this compelling article: Camilla Asplund Ingemark, "The Octopus in the Sewers: An Ancient Legend Analogue," *Journal of Folklore Research* 45, no. 2 (2008): 145–70.

2 Henry Mayhew, *London Labour and the London Poor,* Vol. 2 (London: Griffin, Bohn, and Company, 1861), 155.

3 This story is recounted in several places, including this colorful compilation: Steve Roud, *London Lore: The Legends and Traditions of the World's Most Vibrant City* (London: Arrow, 2010).

4 See a fun cameo in: David Macaulay, *Underground* (Boston: Houghton Mifflin, 1976).

5 Eric Boodman, "Making Sense of the Great Whip Spider Boom," *Undark,* May 17, 2021, https://undark.org/2021/05/17/the-great-whip-spider-boom/.

6 This anecdote and many others appear here: Corey Kilgannon, "The Truth About Alligators in the Sewers of New York," *New York Times,* February 26, 2020, https://www.nytimes.com/2020/02/26/nyregion/alligators-sewers-new-york.html.

7 Thomas Hynes, "Sewer Alligators of New York City," Zoom lecture at the Brooklyn Public Library, August 10, 2021.

8 Elliot C. Clarke, *Main Drainage Works of the City of Boston (Massachusetts, U.S.A.) 1845-1921* (Boston: Rockwell and Churchill, City Printers, 1885), 160.

9 Jessica Leigh Hester, "Please Stop Stuffing Yorkshire Pudding Down the Drain," *Atlas Obscura,* February 27, 2019, https://www.atlasobscura.com/articles/yorkshire-pudding-sewer-blockage. See also: Rose George, *The Big Necessity: The Unmentionable World of Human Waste and Why It Matters* (New York: Metropolitan Books, 2008), 20.

10 Hester, "Please Stop Stuffing Yorkshire Pudding Down the Drain."

11 Jonathan Watts and Rebecca Smithers, "From Babies' Bums to Fatbergs: How We Fell out of Love With Wet Wipes," *The Guardian*, May 11, 2018, http://www.theguardian.com /environment/2018/may/11/from-babies-bums-to-fatbergs -how-we-fell-out-of-love-with-wet-wipes.

12 Fritz Freuthal, 1948, Improvements in and relating to cleaning or sanitary wiping material, U.K. Patent GB646075A, filed January 17, 1948, and issued November 15, 1950.

13 Hillary Dixler Canavan, "A Brief History of the Wet-Nap, Barbecue Sauce's Worst Nightmare," *Eater*, June 17, 2016, https://www.eater.com/2016/6/17/11936294/wet-nap -inventor.

14 Korin Miller, "Weighing the Pros and Cons of Wet Wipes vs. Traditional Toilet Paper," *Self*, September 14, 2016, https:// www.self.com/story/flushable-wipes-vs-toilet-tissue.

15 Jackie Flynn Mogensen, "The Pandemic Has Made At Least One Thing Clear: It's Time to Get on the Bidet Train, America," *Mother Jones*, May 18, 2020, https:// www.motherjones.com/environment/2020/05/bidets -more-environmentally-friendly-toilet-paper-shortage -coronavirus/.

16 In 2020, the Natural Resources Defense Council, an environmental advocacy nonprofit, issued a scorecard ranking toilet paper brands on their sustainability practices, awarding metaphorical gold stars for relying on recycled pulp or other fibers, and docking points for using freshly harvested trees. More than half of the toilet paper brands the council ranked flunked their test.

17 Carly Mallenbaum, "Do You Need a Bidet? The Butt-Spritzing Toilet Gizmos Are Making a Splash in the U.S.," *USA Today*,

January 18, 2019, https://www.usatoday.com/story/life/2019/01/18/bidet-popularity-us-bathroom-toilet-tushy-kohler-american-standard/2590700002/.

18 Terri Coles, "Let's Be Real: Americans Are Walking Around With Dirty Anuses," *Vice*, January 17, 2017, https://www.vice.com/en/article/xyyqk7/lets-be-real-americans-are-walking-around-with-dirty-anuses.

19 Maria Teresa Hart, "The Bidet's Revival," *The Atlantic*, March 18, 2018, https://www.theatlantic.com/technology/archive/2018/03/the-bidets-revival/555770/.

20 Mallenbaum, "Do You Need a Bidet?"

21 Brittany Frater, "It Took a Pandemic, But the U.S. Is Finally Discovering the Bidet's Brilliance," *The Guardian*, April 14, 2020, http://www.theguardian.com/us-news/2020/apr/14/us-bidet-toilet-paper-sales-coronavirus.

22 Lisa Eadicicco, "Bidet Sales Are Soaring as the Coronavirus Causes Toilet Paper Panic-Buying Frenzies Around the World," *Business Insider*, March 13, 2020, https://www.businessinsider.com/coronavirus-bidet-sales-increase-panic-toilet-paper-shortages-brondell-tushy-2020-3.

23 Amanda Mull, "Why Americans Have Turned to Nesting," *The Atlantic*, November 2020, https://www.theatlantic.com/magazine/archive/2020/11/fluffing-your-own-nest/616469/.

24 *NKBA Design Trends 2021*, National Kitchen & Bath Association, 2020, https://nkba.org/research/nkba-design-trends-2021/.

25 Gerald Porter, Jr., "America's Obsession With Wipes Is Tearing Up Sewer Systems," *CityLab*, March 26, 2021, https://www.bloomberg.com/news/articles/2021-03-26/pandemic-wipes-create-sewer-clogging-fatbergs.

26 Ibid.

27 "Wipe Usage Increase Costs Oakland, Macomb Counties Hundreds of Thousands in Cleanup Costs," WJR, April 5, 2021, https://www.wjr.com/2021/04/05/wipe-usage-increase-costs-oakland-macomb-counties-hundreds-of-thousands-in-cleanup-costs/.

28 Katherine Shaver, "A Nasty Pandemic Problem: More Flushed Wipes Are Clogging Pipes, Sending Sewage Into Homes," *Washington Post*, April 23, 2021, https://www.washingtonpost.com/local/trafficandcommuting/flushable-wipes-clogging-sewers/2021/04/23/5e8bbc82-a2c9-11eb-a774-7b47ceb36ee8_story.html.

29 Gerald D. Miller, 1978, Flushable towelette, US Patent US4258849A, filed January 18, 1978, and issued March 31, 1981.

30 See the following news releases from Thames21: "Record Number of Wet Wipes Found on Thames Foreshore," April 24, 2018, https://www.thames21.org.uk/2018/04/record-number-wet-wipes-found-thames-foreshore/. And also: "23 Thousand Wet Wipes Discovered on Stretch of Thames River Bank," April 1, 2019, https://www.thames21.org.uk/2019/04/23-thousand-wet-wipes-discovered-stretch-thames-river-bank/.

31 Ibid.

32 Mark Blunden, "Millions of Wet Wipes Flushed Into the Thames Causing Plastic Nightmare," *Evening Standard*,

July 4, 2018, https://www.standard.co.uk/news/london/
millions-of-wet-wipes-flushed-into-the-thames-are-causing
-plastic-nightmare-a3878976.html.

33 "23 Thousand Wet Wipes Discovered on Stretch of Thames
River Bank."

34 Christopher Meta vs. Target Corporation and Nice-Pak
Products, Inc., Civil action no. 4:14-CV-0832 (Ohio, 2014).

35 When the lawsuit was settled in 2018, the dad received
$10,000 for his work corralling the class-action suit, and
the other members of the suit walked away with much less:
a $1.35 Target gift card or a new pack of wipes for each
package they purchased within particular window of time—
not enough to pay a plumber, let alone repair any damaged
pipes.

36 International Water Services Flushability Group, "Publicly
Available Specification (PAS) 1: 2020 Criteria for Recognition
as a Flushable Product," December 2020, https://www.iwsfg
.org/iwsfg-flushability-specification/.

37 Diane Peters, "Clogging the System: The Feud Over Flushable
Wipes," *Undark*, December 23, 2019, https://undark.org/2019
/12/23/flushable-wipes/.

38 The National Association of Clean Water Agencies, *The Cost
of Wipes on America's Clean Water Utilities: An Estimate
of Increased Utility Operating Costs* (Washington, D.C.:
NACWA, 2020), https://www.nacwa.org/docs/default-source
/resources---public/govaff-3-cost_of_wipes-1.pdf?sfvrsn
=b535fe61_2.

39 Engrossed Substitute House Bill 2565, 2020, http://lawfilesext
.leg.wa.gov/biennium/2019-20/Pdf/Bills/Session%20Laws/
House/2565-S.SL.pdf?q=20210808205202.

40 Macomb County Office of Public Works, "Wipes Clog Pipes Seeking a Solution," YouTube video, 2:31, May 6, 2020, https://www.youtube.com/watch?v=z5BtCC8BijI.

41 The problem has only gotten worse. In December 2021, a local ABC station reported that the city's 2021 clog-busting pricetag was three times its 2018 total. See Anne Emerson, "More Flushable Wipes Getting Stuck in Sewers Than Ever Before Despite Education, Lawsuit," ABC 4 News, December 22, 2021, https://abcnews4.com/news/local/more-flushable -wipes-getting-stuck-in-sewers-than-ever-before-despite -education-lawsuit.

42 Jessica Leigh Hester, "Meet the Fatbergs," *Atlas Obscura*, November 20, 2018, https://www.atlasobscura.com/articles/ whats-inside-fatbergs.

43 Commissioners of Public Works of the City of Charleston vs. Costco Wholesale Corporation, CVS Health Corporation, Kimberly-Clark Corporation, The Procter & Gamble Company, Target Corporation, Walgreens Boots Alliance, Inc., and Wal-Mart, Inc., civil action No. 2:21-cv-42-RMG, January 6, 2021, https://www.classaction.org/media/ commissioners-of-public-works-of-the-city-of-charleston-v -costco-wholesale-corporation-et-al.pdf.

44 Gerald Porter, Jr., "Kimberly-Clark Moves to Settle 'Flushable' Wipes Suit," *CityLab*, April 26, 2021, https://www.bloomberg .com/news/articles/2021-04-26/kimberly-clark-to-settle -flushable-wipes-suit-with-city-agency.

45 Charleston Water System, "Cottonelle Flushable Wipes Are the Only Wipes Safe for Our Sewer System," news release, April 25, 2021, http://charlestonwater.com/CivicAlerts

.aspx?AID=181&fbclid=IwAR1U6hN5yYW40FB_f2DQ6J
_zau1RJx_PlSXhHBQcNV3AHd0bperLsSUVVxE.

46 Jessica Leigh Hester, "Fed Up With Fatbergs, an Australian
Water Company Recorded a Jazzy Jingle About the Sewer,"
Atlas Obscura, August 1, 2019, https://www.atlasobscura.com/
articles/fatbergs-australia.

3 FATBERGS

When wipes, oils, and fats overwhelm pipes, how do you yank them out and get things moving again? And what can we learn from the gunk?

Most of the people wandering the blocks of London's Covent Garden neighborhood on a chilly evening in September 2019, nursing frothy beverages in the gold glow beaming from café windows or hurrying past sleepy tobacconists and hat shops, probably had no idea that the pipes far below their feet were gunked up with globs of fat. They likely had no clue that as they wandered home, a crew was mobilizing for a long, cold night spent extracting the mess beneath the street.

Andy Howard knew something was awry, but he didn't know exactly what to brace himself for: The most reliable way to map the status of the sewer system was to remove a maintenance hole cover and dispatch a camera down to look. In the corners of London that buzz with daytime bustle, the overnight shift is a less-disruptive time to descend into the subterranean network of tunnels that carries wastewater around the city. So, around 10:00 p.m., Howard and his crew started setting up on a brick patch near the posh corner of Pall Mall and St. James, next to a sign pointing visitors toward Buckingham Palace.

Leaning against their trucks, Howard and his colleagues used the patch of brick to stage their gear and incubate their plan. While his crew set up blue gates and orange cones to partition themselves from traffic, Howard perched a coffee cup on the dusty cab of his black Toyota Hilux pickup. With one finger, he squiggled in the dust the meandering route of the Thames and pointed out the sewage works built up along its course. There are so many points along the way, he said, where things can go wrong. And when they do, Howard added, "it's monumental."

"The system is 150 years old—it can only take so much abuse," Howard told me. A decade-long sewer veteran, he looks and speaks like Ricky Gervais, but with thick black glasses and less of a sneer. Howard is a technical specialist with Lanes Group, the company that works with the water utility Thames Water to clean nasty blockages that slow or stop the flow. It's his job to remove gobs of oil, clumps of concrete, and other stubborn things that people have introduced into the underground world.

Purging the pipes is a filthy, unrelenting undertaking— one that confronts anyone who attempts it with a reminder that this subterranean, human-made ecosystem is marked and marred by the mundane choices people make as they move through the world above. Howard and his crew tackle the unsavory proof that our habits are etched under our feet, sometimes with expensive, dangerous consequences.

*   *   *

For as long as sewers have snaked beneath streets, their custodians have battled to stop gunk from building up inside them. In the United States, patents for grease traps date to at least 1884; there, people have combated sewer slurries for more than 135 years.[1]

Tens of thousands of miles of sewer pipes stretch beneath London and the surrounding areas. Many date to the Victorian era, and though the network is intricate and impressive, it has long been dogged by problems. In the mid-19th century, the city's sewer system was notoriously foul and prone to springing leaks. In the 1850s, civil engineer Joseph Bazalgette instructed district engineers to check on the state of their existing pipes. He tasked them with cutting some open and measuring the plaque coating the arteries.

In their report, cross-section sketches from these excremental excavations sit alongside markedly grosser ones from a few years before.[2] Dispatches describe a system that constantly breached its limits. Some of the drawings look prettily grainy—speckled, like pointillist paintings; beautiful renditions of what amounts to a river of shit, stippled with gritty debris. Others are marked with dark crescents, suggesting a pileup of slimy silt. In some, fetid liquid bursts from holes like dark lava, spilling down the sides. Tactful engineers, noting with their tidy annotations the dimensions of the pipes and proportion eaten up by thick deposits, didn't comment on any smells. Still, it's easy to imagine that a whiff would have walloped the nose.

New pipes followed soon after The Great Stink, that menace of 1858 that catalyzed enthusiasm for revamped sewers. The arteries of this expanded network were often brick and they tended to be more spacious than their predecessors. Many still wind beneath London today. Howard refers to the largest he contends with as "the big boy," because "it's big enough to drive a bus down." Bazalgette and company planned the nascent system with a growing city in mind, designing the retooled sewers to cater to a metropolis that might swell to as many as 3.45 million residents—well above London's population at the time, but less than half of the number today. Bazalgette might have found that mind-boggling; his estimate was already lapped by his death in 1891, when more than 4.2 million people called London home.[3]

Pipes have a life expectancy; even under ideal conditions, they would eventually give up the ghost, rusted, corroded, or otherwise weakened. But dumping grease and trash introduces additional stressors that hasten their decline.

Joel Ducoste, an environmental engineer at North Carolina State University who studies underground accumulations of fats, oils, and grease, also known as "FOG," has explained that calcium-rich hard water can cause these deposits to saponify and harden, a bit like the scum that builds on the sides of a bathtub.[4] Fatbergs start small and then sprawl, sometimes growing to be longer than an ocean-crossing plane. The largest of these foul clumps stake out so much of the pipe that water can hardly slosh past them. London has seen several fatbergs of titanic heft, including one that weighed 11 tons.[5] The masses grow beneath other English towns, too: In April 2021, Anglian Water announced that a fatberg extracted from Southend-on-Sea, in Essex, surpassed 440,000 pounds—roughly the size of two blue whales—and was made up of a slew of wipes, menstrual products, kitchen utensils, and more.[6]

Stickiness begets stickiness—one goopy thing attracts more of the same—so a wet wipe that snags on a corroded pipe becomes a nucleus around which grease and fats gather. It's theoretically possible to tell what accumulated first and last, but extraction scrambles any hope of decoding this unwanted accretion. To decode the buildups, some researchers have carted bits of fatbergs into their labs to take a closer look.

When crews in Clinton Township, a Michigan community northeast of Detroit, hauled a fatberg to the surface in September 2018, they set some aside for Tracie Baker and Carol Miller. The mass was six feet deep and spanned nearly the whole 11-foot width of the pipe. Bit by bit, the crew calved the blockage into chunks with axes and saws, and fed the pieces into a wet-vac truck. They dolloped two chunks of the fatberg—vividly described by the public works department's press liaison as a "very thick stew"—into 10-gallon aquarium tanks, which traveled to Baker's lab at Detroit's Wayne State University, where she worked as an environmental toxicologist. (She's now at the University of Florida, in Gainesville.)

The fatberg was an unwelcome addition to the pipes, but Baker thought it also presented a serendipitous opportunity: Since few fatbergs have undergone forensic analysis, she figured it would be interesting to dissect one and describe the stuff inside.[7] It was a time-sensitive proposition; if Baker and Miller were going to collect and study the fatberg, they had to get a move on before it was disposed of. Armed with an $80,000 rapid-response grant from the National Science Foundation, the researchers brought it back to the lab and squared off against a smell so pungent that it stung their eyes until they dripped. When the scientists opened the trash bag that held the fatberg on its journey from the wastewater treatment plant, they spotted flies and wiggling worms. Armed with thick rubber gloves and a fume hood, Baker and Miller, a civil and environmental engineer, recruited some students to help slice and tweeze the thing apart.

Once the sample had dried out, the team dug in and identified candy wrappers, mustard packets, tampon applicators, coffee stirrers, needles, plastic tops from soda bottles, and more—"things I wouldn't think of flushing down my toilet, necessarily," Baker said. They also found tons of wipes. "We obviously knew [they] were going

FIGURE 4 A chunk of the local fatberg was also prepared for display at the Michigan Science Center. (Jessica Leigh Hester).

to be in there," Baker added, though it was impossible to say whether or not the interlopers had been labeled "flushable." She's still curious to learn whether wipe-heavy fatbergs end up trapping pharmaceuticals or other chemicals; if the wipes are "like a sponge," she mused, they could absorb a great deal of junk and potentially become "a super-concentrated, toxic mess." The team hasn't studied that question yet. But the current analysis "confirmed some things that we intuitively knew," Miller told me—particularly that fatbergs are a problem of our own making.[8]

To learn more about how humans create fatbergs, other teams have studied the masses at the chemical level. Using gas chromatography, the environmental microbiologist Raffaella Villa, now at De Montfort University, analyzed a slab of London's Whitechapel fatberg to suss out which specific fatty acids mingled in the blob. Her team primarily found palmitic acids, present in palm and olive oil as well as dairy products and soapy dishwashing liquids. They also detected oleic acid (found in olive oil and almond oil), plus other substances found in cocoa butter, Shea butter, and laundry detergent.[9] Because scientists only sample a small chunk of a huge fatberg, any analysis only tells "one bit of the story," Villa reminded me. But since different oils leave different signatures, Villa suggested that a fatberg on one corner might not be chemically identical to one a few blocks away.

London's current sewer system is overloaded, and at any moment, some of the tubes beneath the city are plugged up. Waging war on fatbergs and other clogs currently costs the United Kingdom tens of millions of pounds each year: Thames Water alone averages 65,000 unclogging missions annually, Howard said, at a cost of £22 million. Work is underway to increase the sewers' capacity, including debuting the Thames

FIGURE 5 A maintenance hole cover celebrates a vanquished fatberg in London. ("Fatberg manhole cover," Matt Brown, licensed under CC by 2.0, https://www.flickr.com/photos/londonmatt/48802632322/in/photolist-2hmwfkd-2kD8xav).

Tideway, a "super sewer" that will store wastewater that now discharges into the river following especially wet weather. (More on that in Chapter Five.)

But that project is still several years from completion. Meanwhile, "We're always on the back foot," Howard told me. "There's never a wait between jobs." Across the city, many happen simultaneously, and don't "stop for anything," Howard added—sewer work continues through weekends, holidays, and pandemics. At 10:20 p.m. on the Tuesday evening I was at Pall Mall, there were 424 clog-fighters on duty across London, and two other workers were on a break. In all, the workers were tackling 75 jobs. (Some are aborted before they even begin: In Shoreditch, one job was postponed because a car was parked on top of the maintenance hole cover that the crew needed to open, and they couldn't find a way in.)

On his phone, Howard showed me the interface where the crew logs their jobs. Each worker had to sign in, confirm that they were feeling mentally fit to descend into the sewer (low on light and high on risks, the enclosed space can provoke claustrophobia and anxiety), certify that they completed the requisite safety checks, and then track their progress. As they go, crews upload photos of the gunk they find. The photos on Howard's phone were almost exclusively of sewer pipes—clean ones, filthy ones, narrow ones, roomy ones. "Most people have pictures of their children," the father of four observed. "I have pictures of sewers." His phone held seven photos of his family and 1,200 photos of the world below the streets.

Crews may begin to suspect that a fatberg looms when water levels go wonky in treatment-plant tanks. The most dramatic hint of a spectacular mass—one of the record-smashing behemoths that make the news—might come when murky waters begin rising in basements or on sidewalks. Another method of fatberg detection is more common and less remarkable: Sewer flushers, who perform routine, constant maintenance on the lines, might notice that water is holding higher than it ought to, suggesting a blockage somewhere—potentially deposits narrowing the room for fluid to flow. Drifting smells offer clues, too. Barry Orr, a graduate student at Toronto Metropolitan University and longtime veteran of Ontario's sewage works whose colleagues call him the "CSI of the sewer system," looks for foam in the sewer line. If his team of flushers sees bubbling, "We know it's grease," he said. Also, he added, "We can smell it."

Howard's crew began the search for the high-water culprit a few days prior, and several blocks away. They started beneath Trafalgar Square—visible from Pall Mall, off in the distance—and then worked their way through the rest of the expanse. They were there to remove mounded oils and trash, as well as silt and concrete that had washed in from road gullies or gathered near construction sites where workers hosed down their equipment without considering how it might harden underground.

Crew members get their bearings with a 10-character code that corresponds to a specific maintenance hole cover. Each is traceable, and understanding the relationship between them is an exercise in infrastructural genealogy. Pairs of letters indicate whether a given sewer is a main branch or an arterial one, and numbers confirm where the section of pipe falls within a given line. At Pall Mall, the crew would explore a trunk sewer dubbed Kings Scholar TS003—the third segment in the Kings Scholar line—accessed through maintenance hole cover TQ29803101. They didn't know what, exactly, awaited them, but it was sure to be gnarly.

*   *   *

To vanquish a fatberg, crews rely on various weapons and techniques. They might hack at the most stubborn ones with pickaxes, for instance, but the first tool is typically water blasted in a hard, swift jet.[10]

The Lanes crew often begins their fatberg fight with a combi, or combination unit, which marries a powerful hose with a strong vacuum. It shoots streams of water from a hose a little more than one inch in diameter at a rate of 124 gallons a minute. The crew might let the water rip three or four times to start, and though they can't really finesse the hose—its aim is somewhat imprecise—the hope is that the brute, bucking force of it will be enough to dislodge the gunk from the pipe's floor and the walls and send it tumbling, so a wider hose can siphon it up. The advantage is that humans can hang out above ground, steering this work by guiding a camera and monitoring the feed in a van parked comfortably and safely on the street. The action is recorded so that the team can review it, poring over eerie, greenish-yellow footage of the sewers' innards.

The jet sometimes works fairly well, especially when the fat is soft and not spackled too thick against the sides of the pipes. When the crew studies footage captured after a successful water blast, they might see something reassuringly mundane—ideally, crisp images of the sewer showing nothing but arching, red-and-ochre bricks. Other times, the jet erodes the buildup just enough to carve out space for a narrow stream, still flanked by solid banks of fat. On the camera footage, that fatberg-choked sewer looks something like a shoveled winter sidewalk, surrounded by dirty mounds of snow in various hues of yellowish gray.

A trickle isn't good enough. Crews want room for water to course—so when a buildup is really stubborn, a person must go down to battle it. This is a dangerous task, and long before London's clog-fighting squad ever drops into a real sewer, they do practice runs in a training facility. There, they hunker down and simulate some of the conditions they might encounter under the streets. They practice quick escapes, rescue maneuvers, and first aid. With a breathing apparatus and an ever-dwindling supply of oxygen, they must find their way through an inky-black maze before the air runs out.

It's an unsettlingly accurate approximation of the real thing and the risks that await there. When you're first lowered into the sewer, you might see a glow seeping in from the street lamp above you. But hook yourself to a tether and walk 300 feet down the line, away from the hole where you entered, and the light vanishes. The subterranean world is cloaked in a thick, velvety black, Howard told me. "Normal darkness is just like closing your eyes in a dark place," he said. "Our kind of darkness has no light penetration at all." If the lights strapped to your uniform are suddenly extinguished, the darkness is overwhelming.

Sewers aren't friendly places for humans. Some tunnels are so skinny that a person can barely squeeze through. Howard described the atmosphere in the pipes as "hostile and unforgiving," lousy with potential foes. A sewer worker can asphyxiate, drown, or wind up trapped in the sticky matrix of gunk. (Crews encounter rats that met unfortunate ends that way, confined in the quagmire.) The gases can kindle fires or fuel explosions. Whoever descends is isolated and maybe submerged in murk, a little like a diver feeling her way beneath waves. The deeper a person descends, the more complicated a rescue would be. Across London, sewer pipes are typically buried between five and 12 meters deep (16 to 39 feet), with anything below eight meters considered especially risky; the Pall Mall sewer is six-and-a-half meters underground.

Because of all the risks, crews do whatever they can to avoid sending people into the sewer. Though many places have piloted other interventions—in Singapore, for instance, some blockages are dislodged by nets or a squeegee—many persistent clogs can only be loosed by hand.

The night before I met them, Howard's crew had blasted almost 100 feet of fatberg from the sewer beneath Pall Mall. They'd hoped that would be enough to restore the flow, but the camera revealed otherwise. I hopped into the passenger seat of their van and watched Howard and his guys scroll through the footage while a woman strolled by after a night out, offering the crew a couple of donuts. (They declined.) Slawomir Punko, a stocky combi operator with a close-cropped beard and a deep reservoir of jokes, wandered to the car to see what he was up against.

Some images were clear. But a few meters away, there was that icky, insidious yellow—proof that the fat hadn't been blasted into oblivion. Someone would have to go in, squatting or kneeling in the muck and using chisels, shovels, or even their hands to pry the fat loose. Physically, the job "does not do you any favors," Howard said. "It does your back in." Now in his fifties, Howard still descends when a complex job demands it. But most of the crew members are between 22 and 32, he figured—and his own days of chiseling away at heavy-duty problems such as gobs of dried concrete are numbered.

The Pall Mall mission would have to wait. The nimble crew members who would descend underground were being summoned for a job that took higher priority: a residential property was flooding with sewage, posing a hazard to life and home. That needed to be addressed first, so the team agreed to pack up and come back later. We fanned out. Howard and I hopped into his car and drove to his next stop of the night, where a young sewer worker was suiting up and getting ready to go deep.

*　*　*

Temperature-wise, sewers aren't terrible. All year long, as poop decomposes, a running sewer holds steady at around 60 degrees Fahrenheit, Howard said. That means "in winter, it is the best place to be," he joked, pleasantly balmier than the streets above. (In summer, he assured me, it's refreshingly cool down there in the gastric juices of the city's belly.)

The tunnels may be temperate, but standing around maintenance hole covers in the middle of the night can induce goosebumps. By the time Howard and I arrived in Greenwich, some seven miles away, the night felt colder. It was tiptoeing toward 1 a.m., and just a few yards away from the Cutty Sark, a 19th-century clipper ship that's now a maritime museum, James Stuart was shivering.

A lanky, beaky guy who now goes by Caspa, Stuart had been sewer spelunking for about a year, ever since he left his job driving a forklift. He descends into the sewer two or three nights a week. (Because the crew rotates positions, he's sometimes up on the surface, working as the 'top man' and running safety checks.) Stuart was already outfitted in waders and protective gear, but the autumn air snuck through the layers. To stay warm, he stomped his feet and flapped his arms, which would soon be tucked into rubber gloves.

Slawomir's brother, Miroslaw, was there as well—and like Slawomir, he had jokes. "We're twin brothers keeping London flowing," he quipped. The pair had joined the company on the same day, six years before.

The crew had been called out to Greenwich a few weeks prior because a nearby basement was dangerously full of gas. The team traced it back to the source, a fatberg in the u-shaped sewer adjacent to the water. Crews had been cleaning it for two weeks, and had only managed to evict a third of the blockage.

Stuart was lowered down. He was gone only a few minutes, walking around and capturing footage for the team to review back up top. When he resurfaced, Howard, Punko, and the rest of the crew gathered around to analyze the images. No one but me seemed to notice that, when Stuart reemerged, his boots were slick with a yellowy-white film, as though he'd trudged through pomade.

In the sewers, fat is not created equal. Some of it has a "mashed-potato consistency," Punko explained. Other fat is buttery, and so yielding that someone walking across it might find themselves sinking. Over time, it hardens until it is almost geological, and solid as rock.

Howard pointed me over to where one of the maintenance hole covers had been pushed aside. I leaned over the partition and listened. I could hear the water rushing, echoing as it barreled along. One of the crew members explained that sound travels fast and far: If someone bangs on a cover a quarter mile down the road, it sounds like it's right next to you. I imagined a river curving through a cave, the sounds of sloshing bouncing off the walls.

Then I sniffed.

Of course, poop doesn't smell great, the team had been telling me, but a fatberg is worse. You can live with the smell of poop, Howard said. Over time, it even starts to smell sweet, he claimed. You get used to it.

No one gets used to the stench of a fatberg.

It's a foul buffet. There's the stink of rotting eggs, courtesy of hydrogen sulfide. Then, something cooked in old, rancid oil. "It's the smell of fries, constantly bombarding you," Howard said. (Maybe—but cold, greasy, and laced with poop.) Gases get trapped beneath a crust on top of the fatberg, Stuart explained. Step too hard and the crust can break, inviting eruptions. Some gases, like methane, are odorless at room temperature. And when other smells are so spectacularly and constantly bad, the nose is no longer a good barometer of danger. Stuart wore a gas meter clipped to the bib of his waders to help him determine when he'd need to bail out.

Punko wandered over to the combi machine and switched it on, while the rest of us ambled over to the open maintenance hole cover and shone flashlights and headlamps, looking to see whether the fat had been shaken loose and was floating past. Nothing much was happening. The stream of water was dark as a moonless river.

That meant the blockage hadn't yet been bested. Work would continue—and the crew would need to stockpile soap and shampoo. Rinsing off the smell of facetime with a fatberg calls for "a lot, a lot, a lot of soap," Stuart said. "The fat gets in your pores," Howard added. "You can smell it for days." Much as our choices linger in the sewer, the smells stick to people who encounter sewage up close. Whenever Stuart emerges from the sewer, he told me, he suds up several times to make sure he has banished the stink.

* * *

Howard put the problem of sewer stewardship to me this way: People think about what kind of light an apartment gets, or what school district it's in, or whether it has a balcony or a yard. They don't ask how old the pipes are or whether there have been any blockages, because, in their minds, the sewer is supposed to be something that just works, no matter what we do to it. We're often asking old pipes to deal with items they were never meant to handle, in a torrent that their designers never anticipated. Howard reminded me that, from London to New York and Singapore, we flush wet wipes, tampons, cement, cooking oil, condoms, and other castoffs that characterize 21st-century urban life, then wonder why our sewers clog to a halt and vomit their contents.

Back at my hotel after I left Stuart and company in Greenwich, I caught a whiff of sulfur as I wiggled out of my coat. The subterranean smell had clung to it, and I hadn't even been close enough to the fatberg to touch it. The stench had drifted up to meet me, and stayed there.

It was a striking contrast to the way that I'd typically experienced sewers. When I considered them at all, it was as a waystation—a place something passed through on its way somewhere else. Even though as a kid I had witnessed the precariousness of our poop infrastructure, I still assumed that the larger systems would just keep working. Despite knowing better, I found myself imagining that everything in the built environment could function regardless of the cumulative consequences of millions of people's daily choices.

Getting closer to the sewer skewered any notion of the pipes as a place where things disappear and nothing lasts. Face-to-face with a fatberg, it is impossible to deny that crud sticks around—in basements, beneath sidewalks and streets, in waterways, on Stuart's skin. Fatbergs are a rebuke—a signal that our habits have consequences, and that "out of sight, out of mind" is a fickle dictum and hollow promise. The sight and smell reminded me that there isn't much distance between the often-hidden infrastructural world and the more obvious one we engage with each day. The boundary is thin, porous, and liable to overflow.

Notes

1 Nathaniel T. Whiting, 1884, Grease-trap, US Patent US306981A, filed September 1883, issued October 21, 1884.

2 Thomas Lovick, Edmund Cooper, George Roe, George Donaldson, and John Grant, *Reports on the Present State of the Pipe Sewers* (London: James Truscott, 1855), London Metropolitan Archives, City of London, MCS/479/024, from the Metropolitan Commission of Sewers collection.

3 Halliday, *The Great Stink of London*, 100.

4 Hester, "Meet the Fatbergs."

5 PA News Agency, "Fatberg Weighing 'More Than an African Elephant' Cleared From London Sewer," *Hampshire Chronicle*, October 29, 2020, https://www.hampshirechronicle.co.uk /news/national/18832364.fatberg-weighing-more-african -elephant-cleared-london-sewer/.

6 "Southend Sewers Cleared of 'Astounding Whale-Sized' Waste," BBC, April 15, 2021, https://www.bbc.com/news/uk -england-essex-56761941.

7 There are a couple notable exceptions, including the mass dissected on the lurid Channel 4 program *Fatberg Autopsy*.

8 Jessica Leigh Hester, "How a Massive Fatberg Went From Sewer to Science Museum," *Atlas Obscura*, January 24, 2020, https://www.atlasobscura.com/articles/detroit-fatberg -exhibition.

9 Hester, "Meet the Fatbergs."

10 Hester, "A Pretty, Seaside Town in England Vanquished Its Giant Fatberg."

4 WATERWAYS

What's traveling from sewers into bodies of water, and why does it matter?

July 1928 in New York City was gummy and miserably hot. In the sun, temperatures reached at least 108, and shade provided only modest relief; anyone lucky enough to find a shadow canopy would still be sweating in 91 degrees.[1] A spate of sizzling days left dozens dead, and as the toll mounted, "Old Man Weather played another cruel joke on us," the *Daily News* reported.[2] Rain blew in and punctured the heat, but ushered in a handful of deaths by drowning.

As temperatures inched up again after the storm, kids in south Brooklyn waded to their ankles or knees in stormwater gathering at soggy street corners. Sewers, overwhelmed by the deluge, were sluggish in carrying the water from the streets. While it lingered, kids stooped and splashed, kicking up sprays of the stagnant water and transforming the pavement into a pool deck or shore.

It was just as well that they didn't trek to the beach—though many of their neighbors did. By one estimate, as many as 3,000,000 New Yorkers fled the confines of their sticky apartments for Coney Island and other swaths of sand.[3] Downpours didn't send them scrambling for cover; many stayed behind and let the drops pelt them, the *Daily News* explained, "happy for real relief that proved only a drop in the bucket."[4]

Within a few days, experts would warn that beaches weren't a safe respite from the stifling streets, and that residents had sewers to blame.

A *Daily News* headline cautioned against "diseased surf."[5] The city's health commissioner, Louis I. Harris, declared the local waters "dangerously unhealthful," and the paper instructed residents to "avoid every bit of water touching Manhattan Island, as well as most of the nearby bays and creeks." Anyone who dared to dip a toe was apparently "gambling with typhoid fever, intestinal diseases of the most serious sort, or eye, ear, nose, or throat ailments." A reported 1.8 billion gallons of sewage rushed into the city's waters each day, and 420 million of those entered the East River, which the paper declared the "most polluted" of all of the city's watery borders. In theory, the *News* noted, the city's waterways should have been residents' "chief refuge from the terrific heat"—but in practice, humans had spoiled them and turned them into something fearsome.

Fouled by a one-two punch of sewage and industrial pollution, New York's waters had sickened and scandalized residents since at least the 19th century. By the late 1880s, Brooklyn residents grumbled that the stench of the sludge-studded, oil-slicked Gowanus Canal offended them during waking hours and jostled them from sleep.[6] In May 1900, the *New-York Daily Tribune* reported on "sewage impregnated water," noting that New Yorkers would "shriek with horror were they asked to drink the unfiltered water of the Hudson River because the sewage of many towns and villages is allowed to run into it."[7] (Unfortunately, the writer added, ice was being harvested from that same contaminated water, so people were still guzzling gross stuff with each cooled-down drink.)

Changes at the federal level wouldn't arrive until the mid-20th century, spurred by another waterway farther west. For years, the Cuyahoga River—which splits Cleveland, Ohio, on its way into Lake Erie—ran so thick with filth that the water sometimes blazed with fire. One such conflagration in June 1969 attracted major national attention, including an article in *Time* magazine, noting that the poor river "oozes rather than flows."[8] The blaze motivated the formation of the Environmental Protection Agency and the passage of the Clean Water Act, two watershed moments for sludgy waterways.[9]

Now, New York City's bodies of water are markedly cleaner, and home to creatures such as the lined seahorse and the oyster toadfish with its perpetually grump-faced grimace. But sewage still ribbons through. So does trash, and other debris that residents pour into their drains, flush down the toilet, or allow to accumulate on streets and sidewalks, where rain washes it into drains.

That's hard to grok. "People are always surprised this is still a problem," explained Stephanie Wear, of The Nature Conservancy, who said that members of the public—and even other scientists with less-crap-filled jobs—raise an eyebrow at her research on sewage-infused waters. "They think these are problems of the past."

They aren't. More than 80 percent of the world's wastewater is unleashed without any treatment, according to a 2017 United Nations report. (And that can be true even if it is centrally corralled: One settlement in the Norwegian Arctic archipelago, for instance, collects sewage in a single pump station the size of an outhouse, and then carries the waste directly into a local fjord via pipes resting on the surface of the cold ground.[10]) Even places that do treat wastewater might not do so as comprehensively as residents may assume: plenty rely only on primary treatment, which entails sifting out the solids that get ensnared in screens or sink to the bottom of a sedimentation tank, a bit like fishing out the floaters and big chunks and leaving the rest. And sewers still routinely overflow. Many are designed to spill when they brim in order to avoid sticky problems such as backing up into people's homes or spewing waste onto the streets.

This tactic is especially common in combined sewer systems, which are less prevalent than they once were. When new systems go in or older ones undergo updates, many are retooled into separated streams, where stormwater and wastewater don't mix.[11] Still, in the United States, sewers in nearly 860 US municipalities are built to belch this way, according to the Environmental Protection Agency, which estimates as many as 75,000 such spit-up incidents a year.[12] These occur even in bustling, wealthy cities that manage to loft residents into gleaming high-rises or shuttle them through labyrinthine underground tunnels. By one count, an annual total of more than 27 billion gallons of sewage and stormwater still plunge into New York Harbor alone.[13]

In New York and elsewhere, from Michigan to Turkey, scientists and advocates are making headway in understanding exactly how much sewage is entering waterways, and what it carries with it, from microplastics and other trash to medications, other drugs, and, of course, poop and the bacteria it carries. Researchers are chronicling what effects this waste introduces when it arrives, and grappling with the question of how to alert residents to the sploshes.

*   *   *

In the summer of 2018, Carrie Roble taught me how to trawl for trash. It was a bright, quiet day, blue above us and calm below—the weather a pleasant 1 on the Beaufort scale, which climbs up to 12 (hurricane-force gales). Armed with a fine-mesh net, we boarded a small boat near Hudson River Park's Pier 40, along Manhattan's western shore, and drifted south toward the Statue of Liberty. Jet skiers shot past, frothing the water. Roble and her team of scientists draped the net over the side of the boat. As we tootled at five knots, it trailed alongside us, skimming the surface.[14]

We were hunting for microplastics, synthetic shards measuring less than five millimeters in diameter. These little pieces—often narrower than a single strand of human hair—can take many forms, from films, lines, and foams to pellets and nurdles, plastics designed to be melted down to mint other plastics.

Humans shed plastic products all over the world. The detritus swirling around the Great Pacific Garbage Patch is infamous; there, wind and waves churn microplastics and their bigger cousins into a long-simmering soup, thick from the surface of the water to the ocean's floor.[15] But plastics—and microplastics in particular—have made their way pretty much everywhere.

Microplastics flutter down across vast landscapes—in 2020, researchers writing in the journal *Science* reported that the equivalent of between around 120 million and 300 million plastic bottles sprinkle protected areas in the American west each year—and also turn up in inland rivers and other freshwater bodies.[16]

Scientists know that microplastics roam around the food web, traveling through the guts of creatures consumed by other creatures (and, eventually, us). Some researchers suspect that they may be associated with metabolic and endocrine issues, but the full picture remains murky. A few animals have learned to live with microplastics, sometimes to their detriment. Tiny caddisflies living in the Brexbach stream in the German town of Bendorf incorporate jagged plastic odds and ends into the cone-shaped cases they live in, so it looks like they have built their houses out of colorful sugar crystals.[17] These cases are less sturdy than the flies' typical digs of densely packed sand, and they're awful camouflage, making the insects vulnerable to hungry trout or dragonflies.[18]

Via plastics, our influence sails and sinks to Earth's most remote locations. Microplastics go low, high—all over the place. A team of Japanese researchers analyzing photos and videos from 5,000 dives over a 30-year period that ended in 2014 found widespread evidence of single-use plastic products in the oceans' deepest nooks. In a paper lyrically titled "Human footprint in the abyss," those scientists described plastics floating thousands of kilometers from shore and lodged more than six-and-a-half miles underwater in the cold, inky reaches of the Mariana Trench.[19] On a recent plastic-sleuthing mission around the Antarctic Peninsula, researchers from Greenpeace and the University of Exeter detected microplastics in each of the eight surface-water samples they collected. Scientists have also found microplastics—from polyester, acrylic, nylon, and more—on Everest, meaning that our microscopic castoffs have ascended the planet's tallest peaks, and stayed behind long after the climbers who carted them there returned home.[20]

Of course, these don't all travel through sewers. Microplastics disperse in all sorts of ways—chipping from a boat's hull, for instance, or sloughing off fishing gear or bobbing trash. But some—such as plastic fibers in wet wipes—do traverse sewer pipes, and some experts believe we've been underestimating the role sewers can play in moving and emitting these little bits of garbage. In London, those artificial landmasses bulging in the Thames—the strips of "Frankenstein foreshore" we met in Chapter 2—are the product of wipes traveling through the pipes, sloshing into the river, and mounding up along London's lip.

A considerable volume of microplastics can also ride along in wastewater itself: Researchers found that the small Arctic village that pumped untreated wastewater into its fjord released roughly the same number of microplastics as a wastewater treatment plant near Vancouver.[21] "Now that it's established that they're pretty much anywhere you look—in water, in air—we're trying to move into, where do they come from, how do they move around, where do they end up?" Julie Dimitrijevic, then a graduate student studying microplastics and blue mussels at Canada's Simon Fraser University, told me in 2018.[22] "We understand that we are putting microplastics in the water, whether through mismanaged waste, or through our wastewater treatment plants," Dimitrijevic said.

Before the US. Congress passed a 2015 ban on microbeads in cosmetics and non-prescription drugs, the little plastic spheres often tumbled down drains whenever people slathered their faces with products such as exfoliating cleansers.[23] (Because the law baked in a several-year cushion for companies to wind down manufacturing and delivery, the miniature menaces circulated for a while after the policy was enacted.) Laundry is another major vector. One count estimates that washing synthetic textiles introduces half a million metric tons of microfibers into the world's oceans each year.[24]

Among microfibers, fleece is particularly fickle. A few years ago, a team of graduate students from the University of California, Santa Barbara examined several jackets made from the fuzzy material. They found that a single trip through the washing machine yielded, on average, around one gram of microplastics—the equivalent of a paperclip or pen cap. (But the machine mattered: Synthetics shed seven times more microfibers in top-loading washing machines, compared to front-loading ones.[25]) A team of Italian researchers described how microplastics twirling in washers and dryers make it to waterways: The fragments are typically small enough to escape being ensnared in the filters at wastewater treatment plants, so they simply sneak through unnoticed.[26]

The folks at Hudson River Park were on the hunt for this type of diminutive detritus—and to find it, they'd have to look closely. Roble, effervescent and outdoorsy, let me tag along as her team sampled water from sites along Hudson River Park, the estuarine sanctuary where she now serves as vice president of the River Project department. As we journeyed down the wet corridor separating New York from New Jersey, the net scooped up samples, sweeping the water for 15 minutes at several sites. When that time had elapsed, we pulled the mesh aboard and squirted it with water to dislodge anything that had snagged. Roble's team rinsed the contents into a plastic container, then poured them into a glass mason jar.

Someone on the boat passed a jar to me. I had never seen Hudson River water up close, and I was mesmerized. In between trawls, I lifted the liquid toward the sun and swirled it like a snow globe. Like one of those souvenir tchotchkes, it seemed to me that the jar contained an entire world. The jar held smears of sand; squishy little jellyfish that looked like globs of clear hair gel; fingernail-sized isopods with scrambling legs; a tangle of rippling brown rockweed.[27] I looked closely, awed by all the life. But close as I looked, there were still things I missed: Someone else had to point out the nearly microscopic hunk of foam on the surface. And the sample was surely studded with plastic too small to notice even if I hadn't been enchanted by the jellies and plants.

To find those plastic bits, the researchers returned to land. There, they poured the contents through a sieve, doused the solids in peroxide to dissolve leaves and other organic material, and then sorted the samples under a microscope, using tweezers to assess each piece. When it was hard to distinguish leaves from green plastic, the team poked suspicious scraps with hot probes to look for melting tendrils.

One purpose of this sampling was to gauge where trash and sewage enters the river and how the refuse mingles with the water. The park spans a four-mile stretch of Manhattan, from Tribeca to Midtown. More than 30 sewer pipes spurt into the park when they're too full, and since 2016, park researchers—in collaboration with scholars at local universities—have sampled the nearby water to learn about how the influx affects the ecosystem.

In 2020, the team published a paper summarizing what they'd learned over several years of testing.[28] After prodding 72 samples at a Brooklyn College laboratory, the scientists concluded that microplastics were widespread—on average, a few hundred thousand clustered in each square kilometer. Roble wasn't surprised, she told me in the spring of 2021. "Microplastics are ubiquitous in water now," she said.

Nor were the researchers shocked to discover that concentrations were often densest close to shore, particularly in rainy years. That could suggest that sewer pipes are superhighways for microplastics. From the get-go, the team suspected that might be the case in New York, because it's true elsewhere: In Paris, for instance, another team of researchers found that combined sewers were the "main and major" culprits routing fragments and fibers into the Seine and Marne rivers.[29] The New York data doesn't quantify how much plastic the sewer outfalls are disgorging, explained Siddhartha Hayes, a study co-author and research and aquaria manager at Hudson River Park, but they do reveal that the environs are hotspots for the pieces.

*   *   *

Of course, plastics aren't the only problem that appears when wastewater spits into ecosystems: The liquid can also carry nutrients, contaminants, and pathogens capable of troubling the environment for the creatures that call it home and the humans who want to enjoy it.

In South Florida, a series of offshore pipes have spent decades feeding billions of gallons of partly treated sewage into the Atlantic Ocean. These outfalls released, as the *South Florida Sun Sentinel* phrased it, "massive clouds that cause the surface to look like it is boiling."[30] One tube in Delray Beach was regularly the pipe du jour for silvery fish that queued up around the soupy green plume, waiting for free meals. The fish seemed satisfied with the arrangement, but the influx of nutrients caused other problems. One professor told the *Sun Sentinel* that the pipes were funneling too much nitrogen and phosphorus into the water and beefing up the growth of algae, which was coating corals. (Reaching for a weirdly appetizing turn of phrase that made my stomach grumble while reading about sewage, the newspaper reporter noted that the corals were "being smothered like hash browns, but it's not in melted cheese."[31]) Happily for Florida corals—and perhaps sadly for the feasting fish—wastewater treatment plants stopped regularly using the Delray Beach pipe in 2009; the others are slated to cease spouting by 2025.

Meanwhile, sewage is also thought to spur the production of a gloopy substance that clogs harbors and curtails sea life around the Mediterranean.[32] In recent years, the coast of Istanbul, Turkey, has routinely been slicked with gunk in umber and dirty-ivory hues. Known as "sea snot"—or, more officially, as marine mucilage—the slime is made up of algae and other microorganisms that trap viruses and other marine microbes. The durable snot can last for months, impeding fishing, disgusting swimmers, and killing corals: In 2007, one research team found that sea snot spread more than 1,550 miles along the Italian coast and stayed there for nearly half a year.[33] In June 2021, when the Sea of Marmara appeared to have been sprayed by a giant's sneeze, Turkish officials vowed to protect the water from sewage and other runoff pollution that boosts nitrogen levels and fuels the spread of the snot.[34] (In the meantime, the government dispatched crews to vacuum the gloop from the water, haul it away, and incinerate it.[35])

Wastewater can also introduce pharmaceutical and illicit drugs, antibiotics, and sometimes even radiographic dyes from medical scans, which dribble out through people's urine.[36] A 2019 study in London found that caffeine, cocaine, and other compounds were most notable in the Thames within a day or two of a sewer overflow, especially when the pipes puked while the tide was out.[37] The concentrations weren't dense enough to be lethal or even especially disorienting to river life—a headline in *The Independent* read, "record cocaine levels in Thames probably not making fish high, experts say"—but there is evidence that fish elsewhere are absorbing some of the pharmaceutical substances into their bodies.[38]

Tracie Baker, the ecotoxicologist, has worked with a large local wastewater treatment plant to study effluent outflows in the Detroit River, a waterway where locals catch walleye and bluegills that they take home and eat. She most recently sampled in the summer of 2021, including at sites near wastewater treatment plants and at outfalls where the contents of combined sewers slosh in. Baker and her team found ultra-persistent chemicals known as PFAS in the surface water, and also in the fish. Baker didn't chronicle the piscine health risks PFAS pose, but in humans, these chemicals—sometimes found in nonstick or stain-resistant products as well as cleaning supplies—can be associated with immune and thyroid problems and cancers. At this point in the data analysis, Baker said, "We can't say for sure that the wastewater effluent is causing this, but we do know that effluent has PFAS in it and so do the fish." Artificial sweeteners turned up around the wastewater treatment plants, too—and because these are produced for (and excreted by) humans, their presence in a natural ecosystem is a reliable sign that wastewater is present there, too.[39] Next, Baker is curious to learn whether PFAS chemicals accumulate on wet wipes, those pesky cloths that linger in sewers or bodies of water.

Another study on accumulated chemicals, this time in the Grand River in Ontario, Canada, found that fathead minnows caged downstream of a wastewater treatment plant took in low concentrations of antidepressants.[40] Pharmaceuticals are also turning up in the blood of silvery bonefish swimming off the South Florida coast. "A yummy crab is a healthy serving" of meds, one scientist recently told journalist Natasha Gilbert, writing in *Hakai Magazine*.[41] (By "healthy," the researcher apparently meant something like "hearty"—scientists found at least 17 different drugs in a single fish.) In South Florida, sewage is sometimes injected deep into limestone as an alternative to sending it straight into the water. But some scientists told Gilbert that it seeps out of the rocks and can eventually wind up there anyway.

In some cases, proximity to a wastewater treatment plant clearly threatens creatures' communities. Researchers working in Texas found that populations of native mussels downstream of treatment plants struggled to survive, probably as a result of being thumped by heavy metals and other materials in the released liquid.[42] Another researcher working in Canada reached similar conclusions, finding that mussels downriver from treatment plants and rinsed by road runoff died younger than their upstream kin, and absorbed ammonia and chloride at levels known to be toxic.[43]

The Hudson River Park crew is keen to know more about what else streams from sewers into the water. Each week from May to October, researchers at the park collaborate with dozens of other organizations on the Citizen's Water Quality Testing Program, a coalition that aims to sniff out pathogens in the river. At upwards of 70 sites, participants collect water samples via kayak or from shore with buckets and poles, and then deliver them to processing locations, including a laboratory at the park. There, the samples are examined for bacteria of the genus *Enterococcus*, which is prominent in poop.

Enterococci are what is known as "fecal indicator bacteria"—a hallmark of something that once wound its way through an animal's digestive tract. Even if these particular bacteria aren't necessarily dangerous to humans, the Environmental Protection Agency explains, they often cuddle up with different bacteria, viruses, and other harder-to-detect microbes that *can* cause disease.[44]

One afternoon in July 2020, Hayes and his colleague Melissa Rex took to Instagram Live to demonstrate how they survey for the bacteria. Wearing blue latex gloves and a black cloth face mask, his dark hair shaved on the sides and pulled back into a stubby ponytail, Hayes prepared a sample that had been collected a few hours prior at Pier 40, where Houston Street approaches the Hudson.

To sleuth out the *Enterococcus*, Hayes and Rex added a packet of powder known as a reagent—"basically just bacteria food," Hayes said—to 90 milliliters of distilled water, and swirled it until the whole thing turned yellowish. Once the sample was uniformly urine-colored, Rex dribbled in 10 milliliters of the Pier 40 water. The team capped and upended the container to prevent the bacteria from settling at the bottom, then poured the sample into a disposable tray with a smattering of little squares. (Picture a translucent keyboard.) After poking it to squash air bubbles and folding it to distribute the water, they slipped it into a sealing machine that looked and whirred like an office copier which heated and sealed the plastic. "It's basically a big laminator," Hayes explained.

The sealed trays spend 24 hours incubating at 41 degrees Celsius (around 106 degrees Fahrenheit), so to demonstrate the rest of the process to the Instagram audience, Hayes relied on sleight-of-hand familiar to anyone who has watched a cooking show where raw batter transforms into a baked, cooled cake: He grabbed a sample that had already cooked. Hayes waved a hand-held UV light across the sample like a metal detectorist combing soil. Any pool that glowed an eerie, winter-moon blue under the UV light was positive for *Enterococcus*. Hayes chuckled a little; over half of the wells shone.

Enterococci levels are measured on the scale of colony-forming units, also called CFUs—essentially, an estimate of the concentration of bacterial cells in a sample of water. The tray holds nearly 100 wells, and any shade of blue is an indication that bacteria is present. To gauge the probable number of CFUs in a particular sample, Hayes and his collaborators consult a chart that matches the number of lit-up cells with an approximation of the bacteria wriggling around in them. In coastal marine waters, the Environmental Protection Agency caps the acceptable limit for *Enterococci* at 35 CFUs per a 100-milliliter sample. (Freshwater limits are slightly lower; the Hudson is a mix of fresh and salty water.) The Hudson River Park team uses three categories, color-coded like a stoplight: Anything at 35 or below is green, meaning safe to swim in or paddle through; samples that test between 35 and 104 are classified as yellow, meaning unacceptable if levels persist; anything above 104 is categorized as dangerous, do-not-go red. Some of the park's water samples have contained 364 CFUs, more than 10 times the recommended limit.

Because they gathered samples over a long period of time, researchers can flag events that coincide with an uptick in those robin's-egg reservoirs. Summarizing several years of findings in a 2020 report, park scientists noted that "fecal contamination spikes" often followed storms, during which untreated sewage gushed into the river.

Storms aren't the only way that *Enterococcus* winds up in the water: The lake in Brooklyn's Prospect Park routinely tests high, Hayes said, "but it's not because of sewage—it's because of geese." The bacteria thrive in birds' guts, too. Human feces can also enter the water without a storm flushing it through the pipes. The Coney Island Creek, for instance, has long been plagued by illegal dumping, including hundreds of thousands of gallons of sewage poured in from more than a dozen buildings for days on end—maybe much longer.[45] But, in the context of known overflows, the bacteria is a reasonable proxy of a spill. The next question is, how do you spread the word that folks should probably stay away?

*   *   *

Stephanie Wear used to live near a North Carolina beach. Over Zoom one cold afternoon, we talked about rain, and the way it sometimes leads to grosser shorelines. "I don't think people connect the dots very well," she said. "They believe that when you flush the toilet, it's taken care of—cleaned, discharged in a way that is safe—but that's just not the case. And whenever there's weather—forget it." Locals knew that beaches often closed after a cloud dump, Wear added, "but people don't connect exactly why." It's often because of sewage.

As a rule, the Hudson River Park team recommends that people avoid the water for two days following a fierce rain, when combined sewers are most likely to expel their haul. Because of the Hudson's swift currents and strong tides, they say, contaminants hang around for 24-48 hours before being pulled out into deeper waters, where they are dispersed and diluted. (That's not to say they're no longer capable of disrupting ecosystems, though, and inlets and other stagnant corners can harbor them longer.) Still, there's much more that researchers want to know. Roble and Hayes see obvious visual evidence of the sewer discharges—little puddles of oil and plumes of waterlogged leaves and trash washed in from the street—up to a day or so after a downpour, but they want more specifics about exactly what's being unleashed. "We have questions around how much water is released, and when, exactly," Roble said. "We would love actual data." And they want to communicate more clearly with residents.

The scientists figure that the residents who use the waterways—who launch boats or cast fishing lines—want those specifics, too. The state gathers some data about pathogens, but their findings offer a zoomed-out, month-by-month view, Hayes said. "It's not as helpful when you want to know, 'Is it safe now? Can I go kayaking this week?'"

Some places make that easier to figure out. In Massachusetts, a bill signed into law in January 2021 will require outfits that maintain outfalls to email or text residents who opt in to notifications within two hours of a detected spill.[46] (The organizations must also alert news outlets and post a public advisory describing the time, date, duration, and estimated volume of the overflow.) In New York, a Twitter bot (@combinedsewer) made by artist and urban planner Neil Freeman does some of that heavy lifting, adapting and sharing city data that would otherwise require a bit of scrolling and clicking.

To keep filling that information gap in New York, Hudson River Park has collaborated with geoscientist Wade McGillis on a publicly accessible digital dashboard that estimates rainfall, sewer discharge, and pathogens over the course of a year. Riffing on another of McGillis's models, this one is based on data gathered at a single site over a rainy 48 hours in the spring of 2020. It extrapolates based on what the scientists know about how pollution spikes and how long it takes to taper, and also pulls from data collected by the United States Geological Survey and the Department of Environmental Conservation. The dashboard, which looks like a calendar, flags days that are a higher or lower risk to water-goers based on predictions about the microbial monsoons rain might trigger. Roble said it's designed to err on the side of overestimating the risk in a bid to keep folks out of the water at times when conditions might be borderline.

To get a firm handle on exactly what's happening where and when, the researchers want granular tracking: sensors at discharge points, for instance, and instruments that measure changes in dissolved oxygen (a proxy for how toxic sewage is making the water for organisms that live there). The humans are keen to know which discharge points are responsible for the most effluent, and also where visitors are entering the water. All would help them prioritize places to intervene.

For the most part, New York's waterways are no longer fetid; the prospect of dipping a toe into the Hudson isn't necessarily horrifying. "Our waterways are cleaner than a lot of people think," Roble told me. "But a big challenge to water quality is our sewers." Making sewers bigger, safer, and more resilient is a heavy lift—one that cities around the world are mulling over as they brace for a crowded, flood-prone future that may push more pipes to the brink.

Notes

1 "Heat Kills 28 in Gotham Sunday," *The Austin Statesman* (Austin, Texas), July 9, 1928, 3.

2 "Rain Brings City Short Relief Only," *Daily News* (New York, New York), July 11, 1928, 2.

3 "Heat Kills 28 in Gotham Sunday," *The Austin Statesman*.

4 "5 Prostrated Fatally, 4 Drowned in Day; Rain a Cruel Joke," *Daily News*, July 11, 1928, 11.

5   "Bathers Warned of Diseased Surf," *Daily News*, July 13, 1928, 16.

6   Jessica Leigh Hester, "Brooklyn's Putrid, Beloved Gowanus Canal Has Been a Horror for Centuries," *Atlas Obscura*, January 14, 2021, https://www.atlasobscura.com/articles/gowanus-canal-superfund-cleanup.

7   "Good for the Health Too," *New-York Daily Tribune* (New York, New York), May 13, 1900, 5.

8   To read more about the 1969 fire and its legacy, see: Jennifer Latson, "The Burning River That Sparked a Revolution," *Time*, June 22, 2015, https://time.com/3921976/cuyahoga-fire/.

9   "HISTORY: Burning River Makes for a Great Story, Comeback Makes for a Greater Future," Northeast Ohio Regional Sewer District on *Medium*, June 23, 2014, https://neorsd.medium.com/history-burning-river-makes-for-a-great-story-comeback-makes-for-a-greater-future-30ec04df9f09.

10 Jessica Leigh Hester, "This Fjord Shows Even Small Populations Create Giant Microfiber Pollution," *New York Times*, September 28, 2021, https://www.nytimes.com/2021/09/28/science/microfiber-pollution-svalbard.html.

11 "Combined Sewage Overflows (CSOs)," RiverKeeper, https://www.riverkeeper.org/campaigns/stop-polluters/sewage-contamination/cso/.

12 "Sanitary Sewer Overflows (SSOs)," United States Environmental Protection Agency, National Pollutant Discharge Elimination System, accessed August 9, 2021, https://www.epa.gov/npdes/sanitary-sewer-overflows-ssos. See also: "Combined Sewer Overflows (CSOs)," United States Environmental Protection Agency, National Pollutant Discharge Elimination System, accessed January 19, 2022, https://www.epa.gov/npdes/combined-sewer-overflows-csos.

13 "Combined Sewage Overflows (CSOs)," Riverkeeper.

14 Portions of this section have been adapted, with permission, from: Jessica Leigh Hester, "Trawling Through the Murky World of Microplastics," *Atlas Obscura*, July 24, 2018, https://www.atlasobscura.com/articles/how-do-you-count -microplastics.

15 "Garbage Patches," National Oceanic and Atmospheric Administration Marine Debris Program, accessed August 9, 2021, https://marinedebris.noaa.gov/info/patch.html.

16 Janice Brahney et al., "Plastic Rain in Protected Areas of the United States," *Science* 368, no. 6496 (June 2020): 1257–1260, https://doi.org/10.1126/science.aaz5819.

17 Sonja M. Ehlers et al., "PVC and PET Microplastics in Caddisfly (*Lepidostoma basale*) Cases Reduce Case Stability," *Environmental Science and Pollution Research* 27 (April 2020): 22380–22389, https://doi.org/10.1007/s11356-020 -08790-5.

18 Jessica Leigh Hester, "These Bugs Armor Themselves in Tiny Shards of Colorful Trash," *Atlas Obscura*, May 13, 2020, https://www.atlasobscura.com/articles/microplastics-cases -caddisflies.

19 Sanae Chiba et al., "Human Footprint in the Abyss: 30 Year Records of Deep-Sea Plastic Debris," *Marine Policy* 96 (October 2018): 204–212, https://doi.org/10.1016/j.marpol .2018.03.022.

20 Freddie Wilkinson, "Microplastics Found Near Everest's Peak, Highest Ever Detected in the World," *National Geographic*, November 20, 2020, https://www.nationalgeographic.com/ environment/article/microplastics-found-near-everests-peak -highest-ever-detected-world-perpetual-planet.

21 Dorte Herzke et al., "Microplastic Fiber Emissions From Wastewater Effluents: Abundance, Transport Behavior and

Exposure Risk for Biota in an Arctic Fjord," *Frontiers in Environmental Science* 9, (June 2021), https://doi.org/10.3389/fenvs.2021.662168.

22 Hester, "Trawling Through the Murky World of Microplastics."

23 "The Microbead-Free Waters Act: FAQs," US Food & Drug Administration, last reviewed February 25, 2022, https://www.fda.gov/cosmetics/cosmetics-laws-regulations/microbead-free-waters-act-faqs.

24 Saabira Chaudhuri, "The Tiny Plastics in Your Clothes Are Becoming a Big Problem," *The Wall Street Journal*, March 7, 2019, https://www.wsj.com/articles/the-tiny-plastics-in-your-clothes-are-becoming-a-big-problem-11551963601.

25 Niko L. Hartline et al., "Microfiber Masses Recovered from Conventional Machine Washing of New or Aged Garments," *Environmental Science & Technology* 50, no. 21 (September 2016): 11532–11538, https://doi.org/10.1021/acs.est.6b03045. See also: "An Update on Microfiber Pollution," Patagonia blog, https://www.patagonia.com/stories/an-update-on-microfiber-pollution/story-31370.html.

26 Francesca De Falco et al., "The Contribution of Washing Processes of Synthetic Clothes to Microplastic Pollution," *Scientific Reports* 9 (April 2019), https://doi.org/10.1038/s41598-019-43023-x.

27 Hester, "Trawling Through the Murky World of Microplastics."

28 Helen Polanco et al., "The Presence and Significance of Microplastics in Surface Water in the Lower Hudson River Estuary 2016-2019: A Research Note," *Marine Pollution*

Bulletin 161, Pt A (December 2020), https://doi.org/10.1016/j
.marpolbul.2020.111702.

29 Rachid Dris, Johnny Gasperi, and Bruno Tassin,
"Sources and Fate of Microplastics in Urban Areas:
A Focus on Paris Megacity," *Freshwater Microplastics:
Emerging Environmental Contaminants? The Handbook of
Environmental Chemistry* 58, ed. Martin Wagner and Scott
Lambert (July 2017): 69–83, https://doi.org/10.1007/978-3
-319-61615-5_4.

30 Ryan Van Velzer, "South Florida Dumps Partially Treated
Human Waste Offshore, But It's Cleaning Up Its Act,"
South Florida Sun Sentinel, April 27, 2017, https://www.sun
-sentinel.com/local/palm-beach/fl-pn-sewage-ocean-outfalls
-20170412-story.html.

31 Bacteria from human guts have also been known to
contribute to white pox disease, an ailment that leaves
elkhorn coral—formally known as *Acropora palmata* and
found across the Bahamas, other parts of the Caribbean,
and Florida—splotchy-looking, as if peeling from an angry
sunburn. See: Kathryn Patterson Sutherland et al., "Human
Sewage Identified as Likely Source of White Pox Disease of
the Threatened Caribbean Elkhorn Coral, *Acropora palmata*,"
Environmental Microbiology 12, no. 5 (May 2010): 1122–1131,
https://doi.org/10.1111/j.1462-2920.2010.02152.x.

32 Jessica Leigh Hester, "What in the World Is Sea Snot?" *Atlas
Obscura*, June 4, 2021, https://www.atlasobscura.com/articles/
what-is-sea-snot.

33 Roberto Danovaro, Serena Fonda Umani, and Antonio
Pusceddu, "Climate Change and the Potential Spreading
of Marine Mucilage and Microbial Pathogens in the

Mediterranean Sea," *PLOS ONE* 4, no. 9 (September 2009): e7006, https://doi.org/10.1371/journal.pone.0007006.

34 "Turkey Says It Will Defeat 'Sea Snot' Outbreak in Marmara Sea," *Reuters*, June 7, 2021, https://www.reuters.com/world /middle-east/turkey-says-it-will-defeat-sea-snot-outbreak -marmara-sea-2021-06-06/.

35 Sarah Zhang, "A Slimy Calamity Is Creeping Across the Sea," *The Atlantic*, June 21, 2021, https://www.theatlantic.com/ science/archive/2021/06/sea-slime-turkey/619256/.

36 Vittoria D'alessio, "Recovering Drugs from Sewers Could Reduce Harm to Wildlife," *Phys.org*, May 27, 2021, https:// phys.org/news/2021-05-recovering-drugs-sewers-wildlife .html.

37 Kelly Munro et al., "Evaluation of Combined Sewer Overflow Impacts on Short-Term Pharmaceutical and Illicit Drug Occurrence in a Heavily Urbanised Tidal River Catchment (London, UK)," *Science of the Total Environment* 657 (March 2019): 1099–1111, https://doi.org/10.1016/j.scitotenv.2018.12 .108.

38 Tim Wyatt, "Record Cocaine Levels in Thames Probably Not Making Fish High, Experts Say," *The Independent*, January 21, 2019, https://www.independent.co.uk/news/uk/home-news /cocaine-london-river-thames-water-research-kings-college -study-fish-high-drugs-a8738146.html.

39 Bridget B. Baker et al., "Persistent Contaminants of Emerging Concern in a Great Lakes Urban-Dominant Watershed," *Journal of Great Lakes Research*, 48, no. 1 (February 2022): 171–182, https://doi.org/10.1016/j.jglr.2021.12.001.

40 Chris D. Metcalfe et al., "Antidepressants and Their Metabolites in Municipal Wastewater, and Downstream

Exposure in an Urban Watershed," *Environmental Toxicology and Chemistry* 29, no. 1 (January 2010): 79–89, https://doi.org/10.1002/etc.27.

41 Natasha Gilbert, "The Search for What's Harming Florida's Beloved Bonefish," *Hakai Magazine*, February 2, 2022, https://hakaimagazine.com/features/the-search-for-whats-harming-floridas-beloved-bonefish/.

42 Trey Nobles and Yixin Zhang, "Survival, Growth and Condition of Freshwater Mussels: Effects of Municipal Wastewater Effluent," *PLOS ONE* 10, no. 6 (June 2015): e0128488, https://doi.org/10.1371/journal.pone.0128488.

43 Patricia L. Gillis, "Cumulative Impacts of Urban Runoff and Municipal Wastewater Effluents on Wild Freshwater Mussels (*Lasmigona costata*)," *Science of the Total Environment* 431 (August 2012): 348–356, https://doi.org/10.1016/j.scitotenv.2012.05.061.

44 "Indicators: Enterococci," United States Environmental Protection Agency, National Aquatic Resource Surveys, https://www.epa.gov/national-aquatic-resource-surveys/indicators-enterococci.

45 Caroline Spivack, "Apartments Dumped 200,000 Gallons of Sewage per Day Into Coney Island Creek," *Brooklyn Paper*, October 4, 2016, https://www.brooklynpaper.com/apartments-dumped-200000-gallons-of-sewage-per-day-into-coney-island-creek/.

46 Bill H.4921, "An Act Promoting Awareness of Sewage Pollution in Public Waters," 191st General Court of the Commonwealth of Massachusetts, January 12, 2021, https://malegislature.gov/Bills/191/H4921.

5 SUPER SEWERS

How do sewers and wastewater treatment plants need to adapt to weather the 21st century?

"If it's noisy down there, we'll give you earplugs," Oliver Smith told me. It was a sunny Wednesday morning in September 2019, and we were walking to a construction site. "I think for the moment, it should be quite quiet," he added. "The guys might be on a break."

Quiet is relative. When I listened to my audio recordings later in my hushed apartment, I heard revving and roaring; it sounded like jackhammers had taken up residence in my ear canals. But Smith might not have noticed the sounds that rattled me; he could be inured to them. His crew works long shifts, and I wondered if the din clung to them the way the smell of the fatberg had woven itself into my garments—if it stayed with them after they'd stowed their tools and returned home.

We were at the Albert Embankment site, where the River Thames ducks under the Vauxhall Bridge in London. We could see the revolving carriages of the London Eye, the wheel that soars over the city. But instead of going up, Smith and I were going down—down flight after flight of stairs to a pit that led into the mouth of a massive sewer pipe.

I stopped on the stairs and craned my neck at the tall walls of cement opening to a disc of sky. I was flanked by scaffolding and a herd of silos holding tons of cement, waiting to be mixed and poured. There were workers, too, in orange uniforms and teal hardhats, waving from a conveyance called a "manrider," a cab that looked like a squat Ferris wheel compartment being lowered into the hole by crane.

We climbed down into the future, and also the past. With his crew, Smith, an engineer on the project, was building one portion of the Thames Tideway Tunnel, a sprawling system that will vastly increase the holding capacity of the city's beleaguered sewers. To do it without disrupting the underground infrastructure that already supports London life, the crew must tunnel through prehistoric ground.

Close to the surface, London's soils are rich with clay. Compressed over millennia, the sediment is strong, but yielding. It's "really good for tunneling," Smith told me. Go deeper, and you'll eventually hit a stratum known as the Lambeth Group. That's "more stony, tougher, a bit more like shale," Smith said. Rockier layers are less cooperative when they meet human tools, and that slows construction.

Eventually, Smith told me, they'd be even deeper, touching ground that "hasn't seen surface for millions of years."

The crews are going deep enough to be humbled, awed, maybe nervously reminded of their bodies, small and fragile in comparison to that ancient dirt. Priests sometimes visit the Tideway dig sites to bless statues of Saint Barbara, the guardian of miners and tunnelers.[1] The figurines stand watch from the wall, overlooking the equipment boring through that long-sunken earth.

*   *   *

In slews of places, wastewater infrastructure renovation projects are struggling or terribly behind schedule. Gibraltar, the British territory on Spain's southern coast, is saddled with a treatment-plant boondoggle years overdue.[2] (One problem is the water; little is desalinated, and residents flush their toilets with water laced with corrosive salt.) In places where sewers and wastewater treatment plants do exist, they're often outdated or under-performing. In March 2021, the American Society of Civil Engineers (ASCE), which issues a regular report card grading the state of the country's infrastructure, gave a D+ to the U.S.'s wastewater treatment plants and 1,300,000 miles of sewers.[3]

The near-flunking grade is no surprise. Many of the country's wastewater treatment systems were built in the 1970s—the era of the Clean Water Act, and the last major investment in sewers—and are wearing out as they age. Most treatment plants were designed to have a lifespan of no more than 50 years, according to the ASCE, which means these facilities have entered their twilight. Some sewers are much older, even approaching supercentenarian status: Close to 20 percent of the pipes that wind stormwater beneath Boston were built before the first World War, reported local radio station WBUR. A share of London's beleaguered pipes has been in the ground since Bazalgette's compatriots laid them there in the 19th century.

The investment that's needed to fix or fend off sewer-related problems often exceeds the city, state, and federal funds earmarked for those overhauls. The ASCE reports that, in 2019, investments in water and wastewater infrastructure fell $81 billion short of the money required for existing issues. The Biden administration's proposed infrastructure plan would offer some assistance for municipal projects that would aim to recover stormwater before it overflows, and also funnel money toward decentralized wastewater treatment systems in individual homes.[4]

Unfortunately, in the United States and elsewhere, aged pipes are increasingly overloaded: In 2021, 15 percent of the country's wastewater treatment plants were intercepting more water than they were designed to encounter, according to the ASCE report. That's partly because many sewer systems are catering to populations much larger than the ones they were built to serve—but it's also because rain is dumping more water into the system.

Scientists expect that as the climate continues to warm, many parts of the planet will be spattered by bursts of fierce rain, even as other swaths weather brutal droughts. (Both are consequences of increased evaporation, according to NASA researchers.) For sewers, climate change will, as WBUR environment reporter Miriam Wasser put it, be akin to "turning up the water pressure."[5] Wetter weather means more water rushing off streets, into catch basins, through pipes, and out into bodies of water—and these deluges are especially damaging when coupled with general sea-level rise.

In tidal areas, many outfall sewers are designed to afford plentiful space for water to slosh out at low tide and then disperse when the water flows back in. That's one process by which bacteria scatter, lowering *E. coli* levels after an overflow. A water level that constantly hangs higher makes it hard for stormwater to spit out into whatever body of water it's designed to enter; it might even cause water to splash backwards in the pipe. "You can't just think about rain, you can't just think about wastewater, you can't just think about sea-level rise," Sanjay Seth, Boston's climate resilience program manager, recently told WBUR—all of those factors are tangled together in one dripping, fetid knot.[6]

To stunt the impact of rain, many cities are trying to store water or slow its movement with the help of green spaces, patchworking the urban grid with rain gardens and other plantings that soak up water that would otherwise hurry into streets.

Other officials are gathering data that can help them anticipate stormwater and wastewater problems, and manage the systems intended to stanch them. These include so-called "smart sewers," outfitted with sensors—often affixed to the underside of the street-facing cover—that track flow, volume, and more. Some places use real-time monitoring to make quick changes (even automatic ones) to the way that sewers or stormwater collection operate.

The city of Hawthorne, California, for instance, installed 50 sensors to track water levels in real-time, helping identify places where daily patterns are fluctuating and where there may be an intruding tree root or a build-up of fats, oils, and grease. Since the sensors were added in 2006, the city has seen a 99 percent reduction in sewer overflows. La Mesa— about 125 miles south, near San Diego—used monitors to scrutinize specific sections of pipe and figure out where the crews needed to target cleaning efforts. Teams had apparently been intervening more than necessary, and were over-cleaning the pipes. (The data led to them nixing 80 percent of their tidy-ups.) Falcon Heights, Minnesota, installed perforated pipes that automatically suck up pond water, making room for stormwater to splash down without flooding a nearby playground or neighborhood.[7] South Bend, Indiana, has become a charismatic poster child for smart sewers. In 2012, the city moved its stormwater management to the cloud.[8] It also debuted automated gates that help control the release of stormwater into the combined sewers. The interventions paid off: Between 2008 and 2014, the city slashed its overflows by 1 billion gallons a year, giving then-mayor Pete Buttigieg something to brag about.

With many sewers pushed beyond their limits, "water companies have no choice" but to occasionally release untreated sewage and stormwater into bodies of water, explained Petra Cox, of England's Crossness Engines Trust.

Such spills are an unpopular solution to a tricky problem. Working with their current infrastructure, water companies are somewhat stuck, Cox said. "They're getting fined, breaking rules, just because there's nothing to be done."

Several places are trying to stave off these problems by scaling up. Chicago—which has a long, choppy relationship with wastewater, and even reversed the flow of its riverine artery to haul sewage away from Lake Michigan, the source of drinking water—is building sprawling holding tanks that resemble lagoons. Singapore is digging deep tunnels. Scaling up is precisely what the Tideway's engineers plan to do. The 15-and-a-half-mile tunnel is designed to intercept some potential overflows and shunt them into holding tanks deep enough underground to dodge subway lines. Instead of disgorging into the Thames, the diverted sewage will travel to a treatment facility. When I visited, the in-progress tunnel looked like a massive bank vault. A rough layer of spray concrete was being covered by a slick inner lining that looked like linoleum flooring. That waterproof membrane—similar to the PVC or resin liners that Singapore is threading into its existing pipes—is an added protection against leakage if cracks splinter the concrete. The slick layer will be covered with a smooth coat of concrete, inhospitable to snags. Backboned by concrete veined with steel fibers, the tunnel is designed to require very little maintenance for several decades, and to safeguard the river for at least a century. The tube yawns nearly 24 feet in diameter; Smith likened it to trading up from a fish tank to a swimming pool, without having to deep-six the existing pipes. Designed with a population of 16 million in mind, it's built to cater to twice as many people as lived in London in 2019 without ever truly filling to the brim.

It's much easier to supersize a sewer's holding capacity than it is to rip out all the existing pipes and install new ones. But a project of this scale becomes a bit of a small city. In addition to the operational tools—the equipment that bores the holes, the lining material, all that cement—the project required bathrooms, changing facilities, lockers, a break room with microwaves, fridges, and long lunch tables. The grounds needed safety protocols, such as color-changing LED lights that line the stairs leading to the pit and turn green when it's safe for someone to proceed down to the site, red when they should halt.

Projects on this scale aren't universally adored. Some critics think the Tideway's £4 billion price tag is excessive, particularly because some of it falls to customers, whose water bills are rising to cover the ballooning cost. (By summer 2021, customers had seen an increase of £18 a year, and some critics expected it to keep inching up to account for the project's delays amid the COVID-19 pandemic.[9])

England's National Audit Office has found that, though the project shaved off several miles of tunneling from the design more than a decade ago, the tunnel could probably still have been shorter and less cavernous, though a narrower tunnel wouldn't have slashed costs the way that a shorter one did, and may have been too slender to do its job.[10]

Meanwhile, some researchers think that the infrastructure calls for different kinds of tinkering, too. Beyond expanding carrying capacity or storage space, they figure that the function of sewers and wastewater treatment plants should evolve in a more expansive way. That may involve making more from the byproducts of poop and other stuff that winds up in the pipes (digesters, for instance, can transform feces into usable biogas, which can power all sorts of things, from stoves to lights).[11] Others believe the priority should be preserving or adding green space to wrangle water before so much of it ends up underground. And still other researchers argue that improving sewers and wastewater treatment plants entails reimagining the functions they can fulfill in the communities they serve. In addition to getting bigger, that idea goes, sewer facilities ought to broaden their scope.

In America, wastewater treatment facilities have cycled through periods of celebration and of desperation, says Simon Betsalel, a recent graduate student in urban placemaking and management at Pratt Institute in New York. For his master's thesis, which he defended in the spring of 2021, Betsalel—who has a background in civil engineering and works as a project manager at the New York City Economic

Development Corporation—traced historical investments in these places and then analyzed several examples to see if wastewater treatment facilities are appealing as public spaces or could be tweaked to work that way.[12]

Betsalel's interest in wastewater dates back 20 years, to when he was an infrastructure-loving kid in Asheville, North Carolina. His class had a chance to visit a wastewater treatment plant, and basically "no one went," he said in a recent presentation about his research; only Betsalel and one other kid showed up. Others "thought it was gross," he said. "Personally, I thought it was awesome. The way this entire system kind of functioned out of sight, out of mind, for most people, and then also just kind of the intrinsic beauty within a facility." He was also impressed by the building's geodesic dome, which looked like something out of Disney's Epcot park.

As a grown-up, Betsalel's research has sprung from his beliefs that built infrastructure is often undervalued, that everyone's life is better with nature in it, and that anything funded by tax dollars should benefit taxpayers in multiple ways. Betsalel wagers that wastewater treatment facilities can do more than funnel and process feces and urine. He believes they can become vibrant spots to walk or read, appreciate art, cultivate plants, and educate the public.

For his thesis, Betsalel studied eight facilities in New York, California, and North Carolina, and found that they had a lot in common. They were typically close to homes or businesses, near water (key for discharging treated liquids), attractive to birds and other wildlife, and had a lot of underutilized space. They were also often cordoned off by large fences and generally closed to visitors.

Betsalel digs the concept of placemaking, the idea that community members can (and should) help shape a space and make it their own. He saw a lot of potential for small changes, such as improving signage so people know what they're looking at, adding murals and other art installations, planting native plants, and promoting open houses or other events that welcome the public in. Then, if funding permits, he said, these smaller changes could lead to larger ones, such as building community spaces within the wastewater campuses. One slide in his presentation put it this way: Wastewater facilities of the future, however big they are, should be "less about shit, more about people."

* * *

On the way to the pit at the Albert Embankment construction site, I passed a green-and-white sign. It held a mirror with a caption that read, "Meet the person responsible for your own safety." I photographed myself smiling into the glassy surface, clutching my notebook in one hand and teal hardhat in the other.

The message made me laugh. It was a playful threat and a clever admonishment: The site had safety procedures in place, but I also needed to keep my wits about me. From holding on to the railings to not wandering in front of the machines, I had a role to play in averting disaster.

Later, the sign struck me as a metaphor for the sewer system itself. There are many measures in place to help keep things flowing, from fatberg-busting crews to pipes big enough to compensate for the ripple effects of our actions, past, present, and future. In one sense, bigger pipes exonerate us, or at least insulate us against the brunt of our behaviors. But the individual choices we make—the wipes we flush, the oil we dump—make a difference, too. We should all be part-time sewer safety wardens. Hard hats optional.

Notes

1 "Priest Blesses Statue 200ft Underground to Protect Tunnel Workers," *Jersey Evening Post*, March 9, 2020, https://jerseyeveningpost.com/news/uk-news/2020/03/09/priest-blesses-statue-200ft-underground-to-protect-tunnel-workers/.

2 Cristina Cavilla, "Sewage Plant Delayed After Company Goes Into Administration," *Gibraltar Chronicle*, July 1, 2020, https://www.chronicle.gi/sewage-plant-delayed-after-company-goes-into-administration/.

3 *2021 Infrastructure Report Card: Wastewater*, American Society of Civil Engineers, https://infrastructurereportcard.org/wp-content/uploads/2020/12/Wastewater-2021.pdf.

4 US Congress, House, *Build Back Better Act*, H.R.5376, 117th Congress (2021-2022), https://www.congress.gov/bill/117th-congress/house-bill/5376/text.

5   Tiziana Dearing, Walter Wuthmann, and Miriam Wasser, "Old Sewers, New Problem: Boston's Stormwater System Is Threatened By Climate Change," WBUR, June 17, 2021, https://www.wbur.org/radioboston/2021/06/17/stormwater-climate-change.

6   Ibid.

7   All of the above examples are drawn from this report: *Smart Data Infrastructure for Wet Weather Control and Decision Support*, United States Environmental Protection Agency Office of Wastewater Management, March 2021, https://www.epa.gov/sites/default/files/2018-08/documents/smart_data_infrastructure_for_wet_weather_control_and_decision_support_-_final_-_august_2018.pdf.

8   Pete Buttigieg, "How South Bend, Indiana Saved $100 Million By Tracking Its Sewers," *Fast Company*, August 5, 2013, https://www.fastcompany.com/3014805/how-south-bend-indiana-saved-100-million-by-tracking-its-sewer.

9   Gill Plimmer, "Thames Tideway Seeks Rise in Water Bills to Cushion London Sewer Delays," *Financial Times*, July 11, 2021, https://www.ft.com/content/f25e29f9-03b4-43a2-9da5-779bcdc3f883.

10 *Review of the Thames Tideway Tunnel,* National Audit Office Department for Environment, Food & Rural Affairs, report by the Comptroller and Auditor General, March 2, 2017, https://www.nao.org.uk/wp-content/uploads/2017/03/Review-of-the-Thames-Tideway-Tunnel.pdf.

11 For much more about this, see: Chelsea Wald, *Pipe Dreams: The Urgent Global Quest to Transform the Toilet* (New York: Avid Reader Press, 2021). Recovering resources from poop also features prominently in Lina Zeldovich's *The Other*

Dark Matter: The Science and Business of Turning Waste Into Wealth and Health (Chicago: University of Chicago Press, 2021).

12 Simon M. Betsalel, "Infrastructure as Public Space: Wastewater Treatment Facilities in Community," master's thesis, Pratt Institute, 2021, ProQuest (28546659).

6 THE INNOVATORS

Experiments are afoot in warehouses and labs, where engineers and other scientists are figuring out how to predict and avoid sewer problems and make the most of whatever can be harvested from the pipes.

Fog hung low in Portland, Maine. The white bodies of seagulls flashed overhead, bright against the pewter sky. It was a little after 8 a.m. on a July morning in 2021, and Jack Chebuske was firing up the truck for his route collecting used cooking oil that would otherwise glug down drains.

Chebuske's name was printed on a green work shirt, which he wore with sleeves rolled up to his elbows. He was sniffly, his blue eyes still sleepy. An iced coffee, his first of two, sweated in a cup holder. "Morning Edition" played on the radio; as we began the day, the 31-year-old explained that he appreciated how the repeating news blurbs doubled as a way to keep track of time.

Chebuske was driving for Maine Standard Biofuels, which corrals used cooking oil and transforms it into other products. The company's drivers sometimes ventured north, nearly to the Canadian border, and south to New York, west of Albany. The day's route was listed on an iPad. The salvage efforts would loop us around Portland for about four hours. Chebuske was circling his hometown, visiting stops he knew well. One was in a grassy backyard, flanked by raspberry bushes, another in a corner of a restaurant kitchen. Others were in alleyways splashed with bright murals, a Black Power fist ringed by yellow sunbeams, a girl with a *calavera* face standing beneath sunflowers and hoisting a sign reading "Abolish ICE." In a few, Chebuske had to join several hoses together and wedge himself between parked cars or crouch under restaurant stairs, where the smell of roasting meat curled in the air.

Some restaurants or bars had hooked up spigots outside their doors from which the oil flowed; they looked like gas or water lines. Most didn't, so again and again, Chebuske flung open lids of plastic bins or metal containers—shaped either like oil drums or BBQ smokers—to reveal the gloop inside. Sometimes it looked like a deep, dark puddle; sometimes rich like brown molasses. Occasionally, pale like amber or ale; once, a soupy green. The thought of it all going down the drain made me feel a little queasy. As I sat to Chebuske's right, bouncing next to him in the cab of the truck, I imagined all the liquid sloshing behind us. Then I tried to picture it in the sewer, where it might not have moved much at all.

Because fats, oils, and grease gunk up pipes, they are unwelcome commuters through the sewer. Intercept or harvest them, though, and they can transform from enemies into useful energy sources. How well that works depends, in part, on where in their life cycle they get captured. Over their tenure in the pipes, fats, oils, and grease have many different morphologies; they start the journey one way, and are transformed by the end.

To figure out the best point in its travels to collect this stuff, Raffaella Villa, the environmental microbiologist at Leicester's De Montfort University, and some colleagues took a close look at samples from homes, restaurants and other food establishments, pumping stations, sewers, and sewage treatment facilities.[1] The scientists scrutinized phlegmy-brown, semi-solid fats, oils, and grease collected from homes, scum from a sewage treatment plant, and scoops from a fatberg itself. They found that the farther the mixture of fats, oils, and grease (FOG) advances through a pipe, the less alluring it becomes for anyone trying to harness it for energy. Samples gathered closer to their sources—such as drains in houses or restaurants—tended to be full of valuable lipids. The choicest oil was the kind that had most recently been sizzling on skillets or gurgling in fryers; farther down the line, the fats and oils had mingled with wastewater and stormwater and become diluted. When it comes to harvesting useful oil, "The fatberg is the worst possible way of collecting it," Villa said. By the time a fatberg forms, its contents are highly contaminated. And before it can be mined for any valuable oils and fats, crews have to dislodge the thing and strip out the rubbish.

Heading off fats, oils, and grease right at the source instead of catching them down the line means enlisting restaurants and other food-service establishments to capture them. Villa's team of researchers estimated that, in an area served by the utility Thames Water, nearly 88,000 tons of FOG could be intercepted at food purveyors, in addition to another 16,000 tons from six million homes. The 48,000 restaurants in the survey area could supply a mass of fats, oils, and grease equivalent to 160 blue whales.

But Villa pointed out that despite the potential energy windfall, it's not always easy to get restaurants and utility companies on board. Collecting these materials involves installing grease traps, which can eat up a lot of space in restaurants where square footage is often at a premium. And in many places, collecting, treating, and transforming the FOG necessitates hopscotching between various utility companies that don't typically work in concert. "If Thames Water collects materials from a pumping station, it's still wastewater, and material that comes out of the digester is still sludge," Villa said. "If you collect material from the restaurant's fryers, it falls under waste legislation." That difference might seem semantic, but it can be meaningful because jurisdiction can influence whether a company sees any reason to push for a change. "At the moment, oil doesn't create a problem for the waste company, so there's no incentive for them to complicate their own life," Villa said.

"The problem is caused to the water company, but the water company doesn't have any say in waste collection."

In some places, FOG is harvested by private companies, instead. Since 2018, residents in the German towns of Erlangen, Fürth, and the district of Roth have been participating in a pilot program to collect cooking oil from homes. People pour their used oil into thick, 1.2-liter yellow-green bottles, and drop them off at vending machines that consume the canisters and spit out empty versions in exchange. The organizer, a company called Altfettrecycling Lesch, picks up the containers, heats the fats to more than 200 degrees Fahrenheit, and swirls them in a centrifuge to separate fats and water from any hitchhiking food scraps. Eventually, the company transforms the processed product into electricity and heat for its own plant, and also fuel for the power grid.

Maine Standard Biofuels collects used cooking oil on an interstate scale and transforms it into biodiesel, which powers vehicles. Biodiesel is produced in all sorts of concentrations, which involve blending the recycled stuff with straight-up petroleum. Most concentrations are B20 or higher—meaning that, at minimum, they're made of 20 percent biodiesel, and 80 percent petroleum diesel, from fossil fuels.[2] According to a tally by the US Department of Energy, 36 states have at least one place to fill up a truck with the recycled fats; Iowa leads the nation in biodiesel fueling stations, with 272.[3] From my pandemic-era perch in Massachusetts, Iowa seemed too far to travel—but I-95 could carry me north to Portland in just over 90 minutes. That's how I wound up riding shotgun with Chebuske. Keen to see how Maine Standard Biofuels pulls it off, I borrowed a gas-drinking car and hit the road.

At each stop, Chebuske flipped a switch inside the truck, uncoiled a hose, and slipped it into the oily slick. The contraption gulped and glugged. "It's basically a big vacuum," Chebuske told me. Sometimes the texture changed by the time the hose reached the bottom—less oil than sludge, and chunky with little bits of food that sometimes looked like charred pork rinds or softening cereal donating its color to darkening milk. When the contraption had slurped the tank dry, Chebuske wiped the oil on his thick gloves, then dabbed them on dirty rags tucked into his pants. The vacuum still whirring, he used the suction to ease the lid closed.

When it was my turn to try slurping the oil, I was surprised by the heaviness of the hose, how it felt almost muscular, like a thrashing snake. When I plunked it into the container and pulled the valve to start the suction, the thing shuddered into action, the hose trumpeting, squealing, emitting a symphony of sounds like—I can think of no polite way to say it—an extremely wet fart. For a minute or so, I held the metal mouth, a nozzle longer than my trunk and legs and known as a "wand" or "stinger," as it bucked and fed oil into the truck's 2,500-gallon tank.

After Chebuske helped me resheath the wand in its holder on the side of the truck, my arms ached as if I had just finished a burst of pushups. This is taxing work. Chebuske, at it for several years, kept loose track of his total oil collected. He said that when he hit a million gallons, a milestone he could almost glimpse through the windshield, he could see wanting to find a job that's a little gentler on the body. "There's only so much slugging it out you can do," he told me, as we drove down a leafy street.

I thought about what I had heard from the crew carving fatbergs from London's sewers. Both types of labor are physically demanding and nearly invisible, necessary and thankless. Chebuske and I encountered one cook at a stop where he siphoned oil from a restaurant basement, but mostly he operated unseen and alone.

Collection fluctuates seasonally, both in terms of the availability of fuel and the work required to suck it up. In the summer, when tourists throng Portland and lobster shacks sling more lunches, some restaurants have hundreds of gallons of oil to hand over, Chebuske said. In the quieter off-season, there's much less oil to go around—and, because the liquid congeals in the cold, whatever there is becomes more stubborn. The oil supply dwindled for a while early in the COVID-19 pandemic, when restaurants were often closed and tourists weren't coming; on the other hand, the threadbare streets made it simpler to maneuver the truck.

Once the oil arrives back at the plant, which sits in an industrial park west of the city, it takes about five days to turn it into something sellable. Freshly offloaded oil sits in heated settling tanks that separate the valuable resource from the wastewater and food particles ("and plastic gloves, straws, and occasionally, rodent bodies," explained Derek Schwartz, the plant manager and engineer). About 20 percent of what Chebuske and the other drivers bring in must be discarded: "Our collection containers look like dumpsters, so people throw trash in them," Schwartz told me. The remaining oil is pre-treated with potassium hydroxide to neutralize the free fatty acids, and then it goes into a reactor with methanol and sodium methylate. Glycerin is removed as it settles out. The oil is then spritzed with water, which drips through and strips out some metals and soaps, then spun in a centrifuge, which draws out the water. What remains is filtered, then heated up to 210 degrees Fahrenheit so that any gases evaporate and are sucked out through a vacuum. Biodiesel falls to the bottom.

The company sells around 500,000 gallons a year, and the plant is a busy place. It's easy to use the soundscape to diagnose problems—"bad clinking, clunking, chunking, all the sounds you hear in your car when you're like, 'Oh God, what's that?'" Schwartz explained, as we stood next to the towering tanks— or be reassured that things are going well. "We use a lot of air tanks; it sounds like the plant has a heartbeat," added Jarmin Kaltsas, the company's founder. "There's a rhythm to the sound of what goes on," Schwartz nodded. "You hear the clinks, you hear the buzzes, everything spins. You can hear liquid pouring onto other liquids, sometimes onto the ground. You've got musical instruments in here."

The finished product powers the company's fleet of trucks, as well as the state's plows and tractors and other vehicles managed by the Department of Transportation, plus some grocery store trucks and ferries that haul people from Portland to the islands freckling the Casco Bay. Anyone can come by the plant to fill up with B99.9 from a single pump out back. When I visited, it was dispensing biodiesel for $2.65 a gallon, a price that had held steady for a year and a half. (And it undercut the less-squelchy competitors: As Chebuske and I wheezed through Portland, I spotted one gas station advertising regular diesel for $3 a gallon.) As fuel prices rose across the board, the team bumped the price to $3.59 a gallon. The company also sells oil for heat, and makes cleaning sprays and soaps from some of the biodiesel byproducts.

Biodiesel has obvious advantages: A gallon emits 80 percent less carbon dioxide than a gallon of petroleum diesel, Schwartz said. And though the oils in biodiesel can seize up in cold weather, the product is thought to be easier on engines, and it spews fewer particulates and no sulfur.[4] But that doesn't mean it currently has wide appeal. "A lot of people have no idea biodiesel exists," Schwartz admitted. (One Irish research team reflected, in 2017, that in the grand scheme of things, diverted FOG is "rarely valorised to bioenergy or biomaterials, despite its potential," though Kaltsas said that, lately, the climbing cost of regular diesel has been revving up customers' interest.[5]) In general, biodiesel doesn't have the crowd-pleasing charisma of other eco-friendly products. Chebuske used to drive for a company in town that collects food scraps to turn into compost, and while elements of the job were similar—he piloted a large vehicle through car-choked streets, stopping to hop out and consolidate the contents of smaller containers into one large one—he noticed a few differences, including people's attitude toward the product. "People just loved it," he said of the compost—they were keen to volunteer, and would share pictures on social media. Used cooking oil is grosser. Though the gig doesn't deter Chebuske from eating at local restaurants—"I'm good at compartmentalizing," he shrugged—it doesn't seem to have quite as many champions. The finished product is useful, but, to some, the process of creating it seems to cross a threshold into the land of *simply too gross*. But a willingness to contend with inherent ick factors is key not only to diverting products from the sewer, but also to pinpointing problems once they're underway.

* * *

It's hard to see inside a sewer. The contents are out of reach, several feet beneath the maintenance hole cover. Entering a sewer involves climbing down a dark, narrow tube; once down there, it's still hard to see the inky soup.

It's a lot easier to scrutinize a sewer's innards when that sewer is inside a lab. So, a handful of universities have brought the sewer—at least, an emulation of it—above ground.

At Toronto Metropolitan University, researchers built a makeshift maintenance hole and shaft that allow them to spy on what happens in a sewer, and find ways to better manage the messes that fester there.

An open-sided cylinder made from pieces of plywood cinched together with metal rings that look like hula hoops, the contraption evokes a really tall dunk tank or a single-person elevator. The bottom holds a trough through which water can flow. The setup is topped with a circular piece of plywood—a faux maintenance hole cover—with a smaller circle cut out. Sometimes, students swaddle the apparatus in thick, dark garbage bags to snuff out the light and approximate the conditions of the real-life pipes.

Anum Khan, a recent graduate student in civil engineering, rigged the faux sewer to survey the water coursing through and the garbage she studs it with. Khan has been searching for a better way for utility companies to detect potential clogs before they form. At the moment, tools such as flow monitors—which measure the water level or how quickly it moves—might reveal that something is amiss, leading crews to suspect that a clog is thriving somewhere along the pipes' route and needs dismantling. "That's a reactive measure," Khan said. "It's: 'This has occurred; we need to go and fix it.'"

Utilities across Canada currently spend enormous sums of money each year on clog clean-ups, and Khan believes they could save a lot by anticipating where the blockages will grow. As part of her graduate work, supervised by Darko Joksimovic, associate professor of civil engineering, Khan is building a system that can document waste in the pipes before it lodges into an unruly clog.

Khan's invention is deceptively simple—essentially a mounted, backlit camera. The component parts are nothing unusual, Khan added, "very cheap technology, off-the-shelf stuff." The novel thing is where Khan is installing it—bracketed on the underside of those maintenance hole covers—and what will happen to the photos. The camera, affixed to a microprocessor the size of cassette tape, will bounce its images to the cloud where an algorithm will screen them for stuff that might become a nuisance. (To ensure the AI is reliably distinguishing between, say, a troubling wipe and a benign wad of toilet paper, an intern will observe over its shoulder for a while.)

The system will allow researchers to flag sites where trash is on the verge of contributing to clogs. It's a way to "identify where problematic locations are in the first place, so you can take a closer look at them," Khan said. Install a camera downstream of a hospital, and you might track latex gloves or disposable masks; surveil a stretch of sewer near a high school, and you could spot tampon applicators or menstrual pads. The idea is to watch many locations over a long stretch of time, Khan explained, and recognize, "This is a problematic location compared to X, Y, Z."

Now that she's done with school, Khan plans to work as the CEO of Knos Technologies (named after the Minoan palace), a company she co-founded to develop and manufacture the product. She has road-tested the invention in the lab by plopping a bunch of products into the rig—wipes, gloves, tampon applicators, and more—and seeing whether the camera picks them up. (When she used light from the visible spectrum to illuminate the darkened field, white objects popped, but blue ones—such as plastic gloves—evaded the camera until she swapped in UV light.) The next step is to sink the gadget underground, and Khan plans to do that soon by partnering with local utilities that manage stretches of narrow pipes that are often occluded.

The dark, dirty sewer environment introduces slews of variables that don't exist in the lab. "Anything that goes into sewers needs to be explosion-proof, gas-proof, waterproof," Khan explained. The module must be tucked into a case that seals it against gases and moisture, because images will be blurry if condensation beads on the lens. Since researchers want the camera to stay in place for a while—three months or so, Khan hopes—the team is tinkering with its power supply.

But that's not to say that all good intel is gained underground. Barry Orr, that longtime sewer worker and current Toronto Metropolitan University graduate student, spends as much time as he can in the school's Flushability Lab. It looks like a hybrid of a warehouse and high-school science lab—exposed pipes snaking across the ceiling; gray, industrial floors. Part of the room is rinsed in purple light to accommodate an unrelated plant-growing experiment. Orr—who has worked as a sewer inspector and outreach coordinator in London, Ontario, since 1995—tries to ignore the lilac ambiance; he's there to study toilets, pipes, and what happens to wipes and other objects that can jam the underground guts.

Orr isn't convinced that tests that rely solely on slosh boxes are persuasive proxies for what goes on below: Those tests produce what Orr calls a "crashing wave" or "roaring falls," and neither gibes with what he's seen in the pipes. "The sewer is like a lazy river ride; it's not a turbulent flow," he said.

Orr wants to know whether wipes and toilet paper disperse in real-time flows that are similar to the kind he encounters in his day job. To figure that out, he uses a makeshift sewer line in the lab.

The tests start with a toilet on a plywood pedestal tiled with linoleum squares. A flush sends anything in the bowl down a pipe and then through a backwater valve (a contraption sometimes installed in new construction to prevent wastewater from regurgitating into people's homes when the sewer is overloaded). The flushed object must pass through several elbows—places where pipes are joined akimbo—and is sometimes visible through cutouts, like the exposed portions of a bobsled track. When the object arrives at the end, Orr catches it in a sieve bucket. Then he subjects it to the slosh box, and sieves it again. The residual pieces are then rinsed and gathered in weighing trays (with crimped sides, they look like aluminum ramekins for creme brûlées). Once dry, the buffeted tailings look like papier-mâché. By comparing the final dried mass to the fresh-from-the-package measurement, Orr can calculate how much the wipe disintegrated.

Orr also wants to see what happens to wipes that have snagged. To simulate that, he uses a pipe in which a gasket was pinched during installation and now protrudes into the tube. Orr snags a wipe on it and runs water at different flows, and for different amounts of time. To show me the result, Orr fished out one wipe that had spent an hour and half getting drubbed by about one liter a second. It was translucent enough to read through—when he splayed it out on the pipe, I could make out "Canada" on the tube beneath it—but clearly intact. The water hadn't wrenched it apart.

Orr performs these tests again and again on scores of products; he has four shelves stuffed with around 150 packages of wipes, paper towels, toilet paper, and other sundries from Japan, Australia, Amsterdam, the UK, and more. (There's also a sock, whose materials—rayon, cotton, polyester—overlap with the contents of some wipes.) He sometimes partners with manufacturers to report on what happens to their products.

Orr is still running tests, juggling commitments at school with his full-time job and spending hours watching highways blur by; the commute from London to school in Toronto is at least two hours each way. He'll continue gathering data to better understand what happens to the things we flush—and whether or not we ought to.

* * *

In addition to diagnosing and quantifying problems in the pipes, researchers are also using labs to explore tactics for defeating the gunk building up inside. At Cranfield University, about an hour northwest of London, England, researchers are building miniature fatbergs and then figuring out how to vanquish them.

Unlike their behemoth, pipe-clogging brethren, these fatbergs fit inside a small beaker. One PhD student makes them with the help of synthetic wastewater. (And also unlike their subterranean kin, these tabletop varieties don't contain wipes or other trash—and they don't reek nearly as much.)

Once the miniature fatberg reaches maturity, researchers subject it to various clog-busting products designed to annihilate FOG, the building blocks of fatbergs. Many products rely on the labor of microbes including bacteria from the genera *Bacillus* or *Pseudomonas*. At wastewater treatment plants, these may keep processes clean and ramp up degradation of solid waste and beef up production of methane, which can in turn be harnessed for energy.[6]

Such bacterial battalions are known as "bioadditives"— biological agents sprinkled into wastewater to increase or enhance microbial communities living in the sewers or plants. "They're like little helpers," said Tolulope Elemo, a chemical engineer at Cranfield. The process of using bioadditives to break down fatbergs is a form of "bioremediation"—the act of deploying biological processes to counteract problems— that can begin in the pipes, before the sewage ever reaches the treatment plant.

It's all too easy to dismiss bioadditives, explained Villa, who has studied them, because their effects can be hard to quantify. There's no single barometer. One way to monitor their work is to measure the fat deposits lining the sides of pipes; another is to assess the oil suspended at the surface of the wastewater. Bioadditives may look potent by one metric, yet puny by another. Gauging whether or not they're working also requires tracking a slew of factors that fluctuate from week to week and neighborhood to neighborhood. How a bioadditive performs at a single moment might have less to do with its inherent capacity for hustle than, for instance, the amount of rainwater flooding the system or trash fouling it up. The ratio of carbon to nitrogen in the waste seems to make a difference, too, at least in one simulation of kitchen wastewater.[7] Skewing the ratio in favor of nitrogen seemed to nudge the organisms to consume fats in the sample, Elemo said, instead of simply filling up on simple sugars and carbohydrates. "It's not one-size-fits-all," Villa noted. "You have to manage the application in the right conditions, in the right place." What works swimmingly in the lab, where flow conditions are simulated, may tank in the sewer, where flows are unpredictable. "We need all the tools in the box to tackle the fatberg problem, and bioadditives are one of them," Villa told me.

At Cranfield, the goal is to see which products fulfill their potential. And once the microbes have proven their mettle in the beakers, scientists introduce them to slightly roomier environs—a pair of 100-meter sewer lines now available for experiments. The lines are fed with real wastewater from a treatment plant on campus. "We have wastewater on tap," said Luca Alibardi, a lecturer in separation processes at the Cranfield Water Science Institute.

Having a pair of pipes is a big advantage, Alibardi added, because it enables researchers to compare a control scenario with an experimental one: They can plop some *Bacillus* into one of the pipes and keep it out of the other. Because the setup is new, the Cranfield crew is currently running wastewater to "condition" the pipes, Alibardi said; a spotless setup wouldn't accurately represent the conditions beneath our feet. It started running in the winter of 2022. Soon, we could be a little closer to slaying fatbergs without quite so much slogging.

The pipes might also help researchers gauge sewers' contributions to greenhouse gas emissions, a primary driver of climate change. Though carbon dioxide is often pegged as the guiltiest party, methane is accused, too, in a recent Intergovernmental Panel on Climate Change (IPCC) report, a nearly 4,000-page primer that thudded into the world in August 2021.[8] (Though some researchers are increasingly convinced that sewers emit this potent gas into the above-ground world at pumping stations and treatment plants, the words "sewer" and "wastewater" appear nowhere in the IPCC document. "Sewer" makes only the briefest of cameos in the several-thousand-page mitigation report that followed in spring 2022—and only in the context of green roofs, parks, urban agriculture, and other tactics for clutching water before it overwhelms the system.[9]) The Ocean Sewage Alliance cites research suggesting that wastewater treatment is responsible for between 3 and 19 percent of all the world's anthropogenic methane emissions, and that the figure would rise if more wastewater winds up treated. Alibardi thinks the control and experimental pipes could be key for utility companies that have pledged to curb emissions within the next decade: In order to meaningfully reduce their output, they must first accurately account for it.

From surveying trash and deploying microbes to trundling through the streets with oil-guzzling machines, scientists and entrepreneurs can help sewers thrive, despite the trash we flush and the other hassles we present to the pipes. Villa put it this way: "There's a massive underestimation of what the sewers could do in addition to what they do." That doesn't let us off the hook for how we treat the sewer, or how the systems are funded and maintained. But it does help draw a blueprint for sewers that work differently—and better.

Notes

1 Thomas Collin et al., "Characterisation and Energy Assessment of Fats, Oils, and Greases (FOG) Waste at Catchment Level," *Waste Management* 103 (February 2020): 399–406, https://doi.org/10.1016/j.wasman.2019.12.040.

2 "Biodiesel Basics," US Department of Energy, Office of Energy Efficiency and Renewable Energy, September 2017, https://afdc.energy.gov/files/u/publication/biodiesel_basics.pdf.

3 "Alternative Fueling Station Counts by State," US Department of Energy, Office of Energy Efficiency and Renewable Energy, Alternative Fuels Data Center, last updated June 7, 2022, https://afdc.energy.gov/stations/states.

4 Daniel Ciolkosz, "What's So Different About Biodiesel Fuel?" Penn State Biomass Energy Center, June 9, 2016, https://extension.psu.edu/whats-so-different-about-biodiesel-fuel.

5 Thomas Wallace et al., "International Evolution of Fat, Oil and Grease (FOG) Waste Management—A Review," *Journal of Environmental Management* 187 (February 2017): 424–435, http://dx.doi.org/10.1016/j.jenvman.2016.11.003.

6 Sai Ge et al., "The Impact of Exogenous Aerobic Bacteria on Sustainable Methane Production Associated With Municipal Solid Waste Biodegradation: Revealed by High-Throughput Sequencing," *Sustainability* 12, no. 5 (February 2020): 1815, https://doi.org/10.3390/su12051815.

7 C. Gurd, R. Villa, and B. Jefferson, "Understanding Why Fat, Oil and Grease (FOG) Bioremediation Can Be Unsuccessful," *Journal of Environmental Management* 267 (August 2020), https://doi.org/10.1016/j.jenvman.2020.110647

8 *Climate Change 2021: The Physical Science Basis*, Working Group I contribution to the Sixth Assessment Report of the Intergovernmental Panel on Climate Change, August 2021, https://www.ipcc.ch/report/ar6/wg1/downloads/report/IPCC_AR6_WGI_Full_Report.pdf.

9 For more on this, see: Yiwen Liu, et al., "Methane Emission From Sewers," *Science of the Total Environment* 524–525 (August 2015): 40–51, https://doi.org/10.1016/j.scitotenv.2015.04.029 and Michael D. Short et al., "Dissolved Methane in the Influent of Three Australian Wastewater Treatment Plants Fed by Gravity Sewers," *Science of the Total Environment* 599–600 (December 2017): 85–93, https://doi.org/10.1016/j.scitotenv.2017.04.152.

7 CONCLUSION

THE AFTERLIFE OF SEWAGE

In 2018, a slab of a fatberg went on view at the Museum of London. It had been dried out, desiccated until its surface was a medley of mottled beiges and browns. The specimen looked almost rocky, with gentle dimples that resembled regmaglypts, the divots that cover some meteorites like a smattering of thumbprints.

I never saw the dried-out slab in person, but I pored over images and videos online, transfixed from the opposite edge of the Atlantic. I couldn't get over the dissonance of it: This thing that looked so natural—that seemed as if it could be made up of iron, silicates, and nickel, the foundations of worlds—was thoroughly synthetic.

My favorite image was one emailed to me by Andy Holbrook, the collections care manager charged with slowly air-drying the sample and ensuring it was safe enough to exhibit. I had reached out to Holbrook because I wanted to know everything I could about cleaning and displaying a fatberg: Why air-dry it? (Placing it in a vacuum chamber would be faster, he explained, but would also contaminate the equipment.) How did the chunk's appearance morph as it went from pipe to vitrine? (The thing shrunk, and the brownish color gave way to a hue Holbrook described as ivory, the color of bone.) Holbrook indulged all my questions, then sent me an X-ray of the fatberg's innards.

The image was a white cloud, splashing across a black background like pooling milk. It was dark in some places and scaldingly bright in others, the fatberg's various densities transformed into a picture that looked like the sky. The image was celestial, captivating, strange. When I interviewed Holbrook about the fatberg for *Atlas Obscura*, he told me he found the X-rays beautiful.[1] "They almost look like nebulae, like pictures from the Hubble Space Telescope," he said, and I agreed. The images astonished me.

On a cloudless night, the dark sky, speckled with beacons, can be sublime, evoking awe and smallness, wonder and fear. Holbrook thought the little bits of glowing grit looked like pinprick stars in a seemingly borderless expanse. I knew I wasn't beholding a stellar nursery, but the view sent some of those feelings of amazement and unease shooting through me, too.

X-rays have long stirred something primal in their human viewers, a mix of fascination and fear. The scan can cast its glance through flesh; it initially seemed like sorcery, revealing something about ourselves we had never seen and were somewhat unnerved to witness. When Anna Bertha Ludwig, wife of X-ray pioneer Wilhelm Röntgen, glimpsed an image of the bones in her hand in 1895, a thick ring wrapping a shadow around a knobby digit, she reportedly said, "I have seen my death."[2]

Ludwig became intimately acquainted with a part of her body she would have never otherwise encountered. The fatberg chunk let visitors see the world inside the sewer, and the X-ray allowed us to peer into the fatberg itself. We tunneled underground and cracked open what we found there.

Once you've gazed inside yourself, you can't unknow the architecture of your body; you can't unsee the scaffolding that props you up. After spending many months thinking about, looking at, and occasionally smelling the inside of sewers, the underground infrastructure often lodges in my head as I follow my routine. I can't unknow what I have learned about these warrens and the many ways we trouble them.

The fatberg exhibition closed years ago, but for most of the time I was writing this book, the museum still broadcast a livestream of the chunk, resting in storage. The camera looked down into the vitrine, and nothing much happened. The fatberg was in the center of the frame, and the edge facing viewers was spiderwebbed with cracks, like parched ground. The thing was still marbled, still dented, still—if you ignored the candy wrapper at the bottom left—plausibly organic. It was almost as if the camera was tracking the decay

FIGURE 6 The fatberg exhibition at the Museum of London was temporary—but the chunk had an additional afterlife in storage. ("Fatberg," Seeing Sanitation, licensed under CC by 2.0, https://www.flickr.com/photos/seeingsanitation/26397173378).

of a downed tree trunk, white crust fungi busy claiming the wood for themselves.

The museum's website offers an overview of the fatberg's post-mortem evolution, observing that over the course of its afterlife on display, it "sweated and changed colour."[3] Things only got worse once it was off view; since arriving in storage, the chunk reportedly "started to grow an unusual and toxic mould, in the form of visible yellow pustules."[4] (The museum team identified it as a member of the bulbous-tipped genus *Aspergillus*, named for its resemblance to an implement used to sprinkle holy water in some Christian churches—though one biologist has written that at least a few "modern taxonomists profess that it more closely resembles a toilet bowl brush."[5])

When I last viewed the livestream, in June 2021, the chunk looked different than I remembered from earlier visits—paler, maybe, and surrounded by more debris than I had recalled. The camera only offered the single overhead view, so I couldn't figure out what, exactly, I was seeing. The castoff bits were small and black, like poppyseeds, and I thought they might be flies; one famously hatched in the vitrine during the initial exhibition, titillating journalists and visitors who took it as further proof that the thing on the pedestal was marvelously putrid. Maybe these were the carcasses of its comrades, I thought—generations of winged lives preserved in front of us. Or maybe they were bits of the fatberg itself, jostled loose when it was moved from display to its new, private accommodations. Maybe they'd been there all along and I simply hadn't noticed.

I stayed on the page for several minutes, letting my eyes clamber from one side of the fatberg to another, studying the terrain. I wasn't sure what I was looking for—the decay to accelerate? The wrapper to flutter in a nonexistent breeze? The flies to perk up and start feasting?—but I felt compelled to stick around and wait.

The stream I was watching had started six months earlier. When I logged on, around 5:30 on a Saturday afternoon, no one else was there—it would have been 10:30 p.m. in London—but the chat reminded me that someone had taken a previous surveillance shift. "anyone here lmao," someone had written. I didn't see a date, and no one responded, but I loved the evidence of another person's lingering. I wondered if they, too, felt stirred and shamed and curious, discomfited and sheepishly unsure of what to do about it.

Caught beneath two layers of plexiglass—less to protect the chunk than to protect visitors from the juices it had stewed in and the gases that might surrounded it—the lump looked distant, an artifact from a past so far gone that it seemed two-dimensional. But the fatberg is contemporary history, and, as Nicola Twilley wrote in an essay about the exhibition for *The New York Times Magazine*, it "functions as a kind of collective self-portrait."[6] The chunk can be read as a condemnation of a global economy that prioritizes convenience over sustainability, and the customers who buy in. The curators also imagined it as a didactic tool, Twilley added: They hoped "that a face-to-face confrontation with their creation [would] provoke the shock needed to change Londoners' habits." That resonates with me. I do find it shocking. And sobering. But empowering, too.

Microbiologists and engineers are constructing possible solutions to sewer woes. But what can the rest of us do? It seems to me that the answer is both collective and individual. We can elect officials who prioritize infrastructure and act with urgency to address climate change, which will worsen sewer troubles. We can press community boards to build stormwater gardens and use other tactics to sop up water before it rushes underground. We can support companies that produce and supply biodiesel. Those strategies involve putting our votes and wallets behind sustainable sewers. On a personal level, we have to hold ourselves accountable, all the time.

That's a hassle. It's easier not to. Petra Cox, from the Crossness facility in the UK, ventriloquized that thought during a chat about the temptation to flush wet wipes. "You're in a little room. No one else is there, no one can see what you're doing, maybe you think, 'Mine won't count,'" Cox said. "'One little wipe won't make a difference.'" She meant this facetiously. Of course, in aggregate, it will.

It all adds up. Over the course of the COVID-19 pandemic, Raffaella Villa, the environmental microbiologist at Leicester's De Montfort University who studies wastewater and FOGs, amassed a collection of glass jars filled with murky water. Picture old salsa jars and jam jars, their contents now wan, grayish and flecked with pearly little globules. "I had loads," she told me. Villa likes tuna packed in oil, and because this liquid is cruel to pipes, she drains it into a separate container instead of sending it swirling down the sink. Villa typically hauls these containers to the bulk waste facility about 20 minutes from her home, where other neighbors offload garden scraps, busted chairs, or bulky furniture. During the pandemic, though, the facility was shuttered for about a year—and the cans collected in Villa's house. This whirls in my brain as I open canned fish and wonder what to do next.

Being a custodian of the sewer is an all-day task: In the morning, I wrangle my wet coffee grounds; if I use a wipe to sponge myself off after an afternoon hike, I decide where to toss the grubby sheet; in the evening, when I sauté vegetables in the skillet, I figure out how to avoid dribbling the olive oil down the drain. After dinner, I hunch in my building's laundry room, trying to decide if my fleece pullover really needs to churn in the washing machine or if I can eke out a couple more wears. When rain hammers down, I remind myself to steer clear of the shower, lest I overload our combined sewer.

Squatting and wiping aren't mindless tasks anymore, either: Barry Orr, one of the graduate students at Toronto Metropolitan University, told me that even the way we handle toilet paper makes a difference down below.

On a July day in 2021, over FaceTime, he walked me over to his toilet-and-pipe setup in the lab and showed me what he meant. He took 12 squares of toilet paper in two six-square strips. He scrunched the first set in his palm and then dropped the squashed squares into the bowl. He flushed the toilet, then, holding his phone, walked me shakily down the pipes so I could see the bundle journeying along. By the time the clump reached the end, it had broken apart; gauzy and fragile, it had begun to disintegrate. He walked back to the toilet and repeated the process—but this time, folded the squares into a little multi-layered parcel that looked like phyllo dough. The stack didn't clear the pipes with one flush; we waited, waited, waited at the end, expectantly watching the water trickle from the pipe as if looking for a kid to barrel, shrieking, down a waterslide. Nothing tumbled out but water; the setup reminded me of a fountain quietly gurgling in a garden. Orr flushed again. The parcel eventually emerged, and when it did, it was still mostly intact. Folding the paper into a thicker stack makes it more durable—which, in the pipes, is exactly what you don't want. It was obvious, I guess—water permeates a smaller percentage of the total surface area of the folded squares, making it harder to dissolve the sheets—but I hadn't thought of it before. Now, I'll always be a scruncher.

Since learning how fragile sewer systems can be and coming to understand some of the tolls of wreaking havoc on them, I consider taking care of the pipes to be part of my civic duty. The planet is suffused with so much suffering that makes me feel helpless—viruses mutating, armies skirmishing, wildfires and heat waves devastating communities, songbirds dying in a season that should be loud with their music. Against this backdrop, I have started to think about sewer stewardship as an antidote for existential dread. At least I can help the pipes run smoothly and help preserve the ecosystems they touch. At least I can do that.

"It's not just up to the city, it's not just up to the property owner on the waterfront, it's really up to all of us to make this work," Sanjay Seth, the climate resilience program manager for Boston, told WBUR during a segment about future-proofing the city's sewers.[7] I don't think residents should have to pick up the slack for underfunded systems at the end of their infrastructural lives as an alternative to municipal investment in these projects—but I do think we ought to pitch in, or at least pledge to avoid taxing the systems. It could be as simple as setting up a rain barrel. "You could go to Home Depot today and you could figure out how you could have an impact," Seth said. In a world of complex problems, simple solutions—even partial ones—feel like a balm.

I have written most of this book a few miles from Walden Pond, a deep pool carved by an ancient glacier that heaved across land that is now Massachusetts. In the 1850s, the writer and naturalist Henry David Thoreau famously camped out in a small cabin near the pond's blue-green shore. Surrounded by slippery, muscular fish, lounging turtles, acrobatic birds, and a bevy of birches, pines, and other trees he once described as fringing the water like "slender eyelashes," he stared into the pond and mused about what it showed him.[8] "Thoreau . . . recognized that Walden could be both a window and a mirror," the geoscientist Curt Stager, who has studied Walden's sediments, wrote in the *New York Times* in 2017.[9] "He called it 'Earth's eye,' in which we can see ourselves and our world reflected." I often visit the pond and stand at the lip of the shore. I see my face held in the water until darting fish blur my eyes or smear my mouth.

Sewer muck isn't literally like that; a fatberg won't reflect your face back to you. But I keep thinking about what sewers reveal about the way people treat the planet, and how we might be gentler. We often meddle with the world in ways that make it harder for many species, including our own, to survive here. Our choices accrete, even the small ones, even the ones that seem like they can't possibly matter; the pipes, water, and land are laden with the weight of them. Maybe it's apt to think of all our cumulative habits as water, rushing through a pipe—ever mingling, never alone.

Notes

1 Jessica Leigh Hester, "How a Museum Cares for a Giant Chunk of Dried-Out Fat," *Atlas Obscura*, February 16, 2018, https://www.atlasobscura.com/articles/london-fatberg-museum-exhibition-conservation.

2 Howard Markel, "'I Have Seen My Death': How the World Discovered the X-Ray," *PBS Newshour*, December 20, 2012, https://www.pbs.org/newshour/health/i-have-seen-my-death-how-the-world-discovered-the-x-ray.

3 Museum of London, "FatCam! Watch Fatberg in Real Time," news release, August 14, 2018, https://www.museumoflondon.org.uk/discover/fatcam-watch-fatberg-live.

4 Ibid.

5 Antonis Rokas, "*Aspergillus*," *Current Biology* 23, no. 5 (March 2013): R187–R188, https://www.sciencedirect.com/science/article/pii/S0960982213000249.

6 Nicola Twilley, "Letter of Recommendation: Fatbergs," *The New York Times Magazine*, March 27, 2018, https://www.nytimes.com/2018/03/27/magazine/letter-of-recommendation-fatbergs.html.

7 Dearing, Wuthmann, and Wasser, "Old Sewers, New Problem: Boston's Stormwater System Is Threatened By Climate Change."

8 Henry David Thoreau, *Walden: Or, Life in the Woods*, 1854, Project Gutenberg eBook #205, https://www.gutenberg.org/files/205/205-h/205-h.htm.

9 Curt Stager, "Opinion: What the Muck of Walden Pond Tells Us About Our Planet," *New York Times*, January 7, 2017, https://www.nytimes.com/2017/01/07/opinion/sunday/what-the-muck-of-walden-pond-tells-us-about-our-planet.html.

ACKNOWLEDGMENTS

This project is the product of many intrepid and steel-stomached supporters. Thanks to Monika Woods for her patience and imagination. Thanks to Haaris Naqvi, Christopher Schaberg, and Ian Bogost at Object Lessons for their willingness to be repulsed, to Rachel Moore and the production team for logistical wizardry, to the copyeditors for finesse, and to Alice Marwick for the dreamy cover. At *Atlas Obscura*, I'm grateful to Samir Patel for indulging my love of unsavory stories (from which some of this text was adapted) and to Sommer Mathis, Daniel Gross, Jeremy Berlin, Lex Berko, Ella Morton, Gemma Tarlach, and April White for sharpening my ideas and being so generous with their own. Thanks to Samantha Chong for reading many of these chapters with affection for gross details, and to Michelle Cassidy for brilliant tweaks. Tremendous gratitude to David Dudley, Amanda Kolson Hurley, Nicole Flatow, Kate Julian, and Jane Nussbaum for the gift of improving my work. Cheers to my many wonderful former coworkers, including Laura Bliss, Cara Giaimo, Sarah Laskow, Chris Naka, and Sabrina Imbler for their camaraderie, integrity, and beautiful stories.

Thanks to my colleagues at Johns Hopkins for their kindness as I finished this manuscript. Deep gratitude to Matt Taub for helping to shoulder the fact-checking.

Thank you to all who shared their expertise, especially everyone who put up with my pestering and let me join them as they worked. (And the Toronto Metropolitan University crew, who brought me along via FaceTime!) This book would be flatter without you, and I'm so grateful for your time. I'm also grateful to the London Metropolitan Archives and to the staff at the small-but-mighty Lincoln Public Library in Massachusetts for keeping the books coming in during a global pandemic.

Thank you to my MFA teachers at Hunter—especially Saïd Sayrafizadeh, advisor extraordinaire—and to my talented and eagle-eyed classmates, including Kate Farrell, Wendy Hammond, Jonathan Rizzo, and Liz Palombo, who read several of these chapters and lobbed great questions and suggestions.

This book is for my grandparents, Nancy and Lloyd Robertson, whose faith in me has the staying power of a fatberg. To the family that raised me—Mom, Dad, Cam, Nonnie, Dad-Dad, the whole Robertson clan—thank you for reading to me, bringing me to museums and nature centers, and teaching me to be curious about the world. I'm lucky to receive your bottomless enthusiasm, encouragement, and good humor, and deeply grateful to be yours.

I am also awed by my chosen families. It's an honor to have earned the friendship of the luminous and wild-smart Ginny Rangos-Burridge, Angela Chen, Sofia Gans,

Jessica Glazer, Maryn Liles, Sara Raftery, Steph Strauss, and Lyndel Sorenson, all of whom buoyed me throughout the writing. Many other dear pals and colleagues near and far shared pep talks, offered proofreads, and listened to a litany of disgusting facts. Thanks to Alex Mayyasi for being this book's first champion and sharing smart notes and a well-stocked snack drawer, and to Christina Djossa for thoughtful pick-me-ups when I felt stuck. Thank you to the extended Creel and Nichols crew—Kathleen, Katie, Buck, Elinor, Olivia, and both Leslies—for welcoming me with wide-open hearts, minds, books, and refrigerators, and indulging sewer talk at dinner for an entire year. To RCC, my love, trusted reader, and favorite scientist: Thank you for your slaughtered adverbs, your spectacular brain, and your magnificent spirit. I'm so glad to wade together through the muck.

INDEX